AF356140

Energy Harvesters and Self-Powered Sensors for Smart Electronics, 2nd Edition

Energy Harvesters and Self-Powered Sensors for Smart Electronics, 2nd Edition

Editors

Qiongfeng Shi
Huicong Liu

Basel • Beijing • Wuhan • Barcelona • Belgrade • Novi Sad • Cluj • Manchester

Editors
Qiongfeng Shi
School of Mechanical and
Electronic Engineering
Southeast University
Nanjing
China

Huicong Liu
School of Electric Science
and Engineering
Soochow University
Suzhou
China

Editorial Office
MDPI
St. Alban-Anlage 66
4052 Basel, Switzerland

This is a reprint of articles from the Special Issue published online in the open access journal *Micromachines* (ISSN 2072-666X) (available at: https://www.mdpi.com/journal/micromachines/special_issues/Energy_Harvesters_and_Self_powered_Sensors_II).

For citation purposes, cite each article independently as indicated on the article page online and as indicated below:

Lastname, A.A.; Lastname, B.B. Article Title. *Journal Name* **Year**, *Volume Number*, Page Range.

ISBN 978-3-7258-0158-9 (Hbk)
ISBN 978-3-7258-0157-2 (PDF)
doi.org/10.3390/books978-3-7258-0157-2

Contents

About the Editors . . . . **vii**

Preface . . . . **ix**

Qiongfeng Shi and Huicong Liu
Energy Harvesters and Self-Powered Sensors for Smart Electronics, 2nd Edition
Reprinted from: *Micromachines* **2024**, *15*, 99, doi:10.3390/mi15010099 . . . . **1**

Ning Li, Hu Xia, Chun Yang, Tao Luo and Lifeng Qin
Investigation of a Novel Ultra-Low-Frequency Rotational Energy Harvester Based on a
Double-Frequency Up-Conversion Mechanism
Reprinted from: *Micromachines* **2023**, *14*, 1645, doi:10.3390/mi14081645 . . . . **6**

Long Zhang, Hang Shao, Jiaxiang Zhang, Deping Liu, Kean C. Aw and Yufeng Su
Improvement of the Airflow Energy Harvester Based on the New Diamagnetic Levitation
Structure
Reprinted from: *Micromachines* **2023**, *14*, 1374, doi:10.3390/coatings11030274 . . . . **24**

**Yizhi Li, Jagadheswaran Rajendran, Selvakumar Mariappan, Arvind Singh Rawat, Sofiyah
Sal Hamid, Narendra Kumar, et al.**
CMOS Radio Frequency Energy Harvester (RFEH) with Fully On-Chip Tunable Voltage-Booster
for Wideband Sensitivity Enhancement
Reprinted from: *Micromachines* **2023**, *14*, 392, doi:10.3390/mi14020392 . . . . **41**

**Balamahesn Poongan, Jagadheswaran Rajendran, Li Yizhi, Selvakumar Mariappan,
Pharveen Parameswaran, Narendra Kumar, et al.**
A 53-µA-Quiescent 400-mA Load Demultiplexer Based CMOS Multi-Voltage Domain Low
Dropout Regulator for RF Energy Harvester
Reprinted from: *Micromachines* **2023**, *14*, 379, doi:10.3390/mi14020379 . . . . **58**

**Wasim Dipon, Bryan Gamboa, Maximilian Estrada, William Paul Flynn, Ruyan Guo and
Amar Bhalla**
Self-Sustainable IoT-Based Remote Sensing Powered by Energy Harvesting Using Stacked
Piezoelectric Transducer and Thermoelectric Generator
Reprinted from: *Micromachines* **2023**, *14*, 1428, doi:10.3390/mi14071428 . . . . **73**

Qi Wang, Tao Ruan, Qingda Xu, Zhiyong Hu, Bin Yang, Minmin You, et al.
A Piezoelectric MEMS Speaker with a Combined Function of a Silent Alarm
Reprinted from: *Micromachines* **2023**, *14*, 702, doi:10.3390/mi14030702 . . . . **87**

**Stalin Allwin Devaraj, Kambatty Bojan Gurumoorthy, Pradeep Kumar, Wilson Stalin Jacob,
Prince Jenifer Darling Rosita and Tanweer Ali**
Cluster-ID-Based Throughput Improvement in Cognitive Radio Networks for 5G and
Beyond-5G IoT Applications
Reprinted from: *Micromachines* **2022**, *13*, 1414, doi:10.3390/ mi13091414 . . . . **98**

Yuting Zhang, Qianhui Ge and Yanhan Zeng
A 0.6 V_{IN} 100 mV Dropout Capacitor-Less LDO with 220 nA I_Q for Energy Harvesting System
Reprinted from: *Micromachines* **2023**, *14*, 998, doi:10.3390/mi14050998 . . . . **108**

Chuting Wu, Jiabao Zhang, Yuting Zhang and Yanhan Zeng
A 7.5-mV Input and 88%-Efficiency Single-Inductor Boost Converter with Self-Startup and
MPPT for Thermoelectric Energy Harvesting
Reprinted from: *Micromachines* **2023**, *14*, 60, doi:10.3390/mi14010060 . . . . **125**

Qian Lian, Peiqing Han and Niansong Mei
A Review of Converter Circuits for Ambient Micro Energy Harvesting
Reprinted from: *Micromachines* **2022**, *13*, 2222, doi:10.3390/mi13122222 . . . . **140**

About the Editors

Qiongfeng Shi

Qiongfeng Shi is currently a Professor at the School of Electronic Science and Engineering, Southeast University, Nanjing, China. He received his B.Eng. degree from the Department of Electronic Engineering and Information Science at the University of Science and Technology of China (USTC) in 2012, and received his Ph.D. degree from the Department of Electrical and Computer Engineering at the National University of Singapore (NUS) in 2018. He then worked as a Research Fellow in the same department at NUS from 2018 to 2022. His research interests include flexible/wearable electronics, energy harvesting, multimodal sensing, human–machine interfaces, and intelligent systems. He has authored and co-authored more than 100 high-quality journal papers and conference papers in his research fields, which have received more than 8000 citations with an H-index of 50 and an i10-index of 78. He has served as the Young Advisory Board Member for InfoMat and InfoScience, the Guest Editor for several international journals (including *JMM*, *Nanomaterials*, and *Micromachines*), as well as a reviewer for more than 40 peer-reviewed journals.

Huicong Liu

Huicong Liu is currently a Professor at the Robotics and Microsystems Center, School of Mechanical and Electrical Engineering, Soochow University, China. She received her B.Eng. and M.Sc. degrees from the Department of Mechanical Engineering, University of Science and Technology Beijing, China, in 2006 and 2008, respectively, and the Ph.D. degree from the Department of Mechanical Engineering, National University of Singapore (NUS), in 2013. She was a Research Fellow with the Department of Electrical and Computer Engineering, NUS, from 2012 to 2013. Her research interests are vibration-based MEMS/NEMS energy harvesters, self-powered MEMS/NEMS systems for IoTs, and flexible and intelligent sensors used for human–machine interfaces. Prof. Liu has contributed 2 academic books and more than 100 academic papers. Her work has been cited more than 4100 times with an H-index of 31. She was the co-chairman of the 2nd International Conference on Vibration Energy Harvesting (VEH 2019), the TPC member of IEEE NEMS 2018-2019 and 2021, and the TPC member of Transducers 2021.

Preface

Over the past year, we have seen continuous rapid advancements made in the field of energy harvesting and self-powered sensing technologies, enabling the realization of the Internet of Things (IoT) and the Artificial Intelligence of Things (AIoT) in various applications such as personalized point-of-care networks, smart homes, unmanned shops, smart manufacturing, smart transportation, digital twin, metaverse, etc. By contintuing with the 2nd edition of the Special Issue "Energy Harvesters and Self-Powered Sensors for Smart Electronics", we aim to include and showcase new important progress in this field, as well as provide readers with unique insights into energy harvesting mechanisms, materials, devices, circuits, and systems.

In this reprint, we compile nine research articles and one review article along with an Editorial written by us. These high-quality articles from researchers across the world showcase the latest advances in mechanical energy harvesters, radiofrequency energy harvesters, self-sustained IoT sensors, power management, and functional systems in a broad range of IoT applications. Based on ongoing progress, we can see two clear trends in this field: one is to continue increasing the device performance with higher energy conversion efficiency, higher sensitivity, and better adaptability, and the other is to integrate the system with more functionality and higher intelligence. From research on commercialization in practical scenarios, good system reliability and robustness also need to be considered and designed carefully.

In the end, we want to thank all the authors for their valuable contributions to this reprint, without which it would not be possible for us to compile this reprint. We hope that the articles included here are helpful and inspiring for readers who want to learn more about the research advances and technical insights within the field. With continuous efforts and innovations, we believe that more and more self-sustained devices and systems will soon enter, and play important roles in, our daily lives, enabling more sustainable development within our society.

Qiongfeng Shi and Huicong Liu
Editors

micromachines

MDPI

Editorial

Energy Harvesters and Self-Powered Sensors for Smart Electronics, 2nd Edition

Qiongfeng Shi [1,*] and Huicong Liu [2,*]

[1] Interdisciplinary Research Center, School of Electronic Science and Engineering, Southeast University, Nanjing 211189, China
[2] School of Mechanical and Electrical Engineering, Jiangsu Provincial Key Laboratory of Advanced Robotics, Soochow University, Suzhou 215123, China
* Correspondence: qiongfeng@seu.edu.cn (Q.S.); hcliu078@suda.edu.cn (H.L.)

Citation: Shi, Q.; Liu, H. Energy Harvesters and Self-Powered Sensors for Smart Electronics, 2nd Edition. *Micromachines* **2024**, *15*, 99. https://doi.org/10.3390/mi15010099

Received: 29 December 2023
Accepted: 30 December 2023
Published: 4 January 2024

1. Introduction

With the worldwide rollout of the 5G communication network and 6G around the corner, we have witnessed the rapid development of the Internet of Things (IoT) technology, enabling big data and digital transformation in various fields [1], e.g., remote healthcare, intelligent body area network, smart farming, smart building, industry 4.0 and smart transportation. The successful implementation and functionality of IoT networks require support from an enormous number of interconnected sensor nodes for localized and holistic information acquisition. The conventional power supplying strategy using batteries possesses inevitable drawbacks of a limited lifespan, bulky/heavy characteristics, environmental unfriendliness and difficulty in replacement, greatly hindering the sustainable development of IoT networks. In this regard, energy harvesting technologies that can scavenge available energy (e.g., light, thermal, kinetic and radio frequency energy) from the surroundings appear to be promising candidates in combination with batteries for sustainable power supply in the IoT era [2]. The widely adopted energy harvesting technologies include piezoelectric, triboelectric and electromagnetic mechanisms for kinetic energy [3–5]; thermoelectric and pyroelectric mechanisms for thermal energy [6,7]; photovoltaic effect for light energy [8]; and antenna rectification for radio waves [9]. In practical implementation, energy harvesting devices can be configured into energy harvesters for continuous power feeding or self-powered sensors for zero-powered sensing [10], both contributing to higher sustainability in standalone IoT sensor nodes. The research field is currently focused on achieving output performance improvement, efficient power management and compact system integration via mechanism, material, structure and circuit innovation, paving the way toward functional and self-sustainable IoT nodes and systems.

This Special Issue is the second edition of "Energy Harvesters and Self-Powered Sensors for Smart Electronics", and it showcases the continuous and important progress in the field of energy harvesting and self-powered sensing. This edition contains 10 published articles (9 research articles and 1 review article) that address the urgent demand of sustainable power supply for IoT systems, aiming to further advance the field through detailed explorations of device design, circuit optimization, strategy enhancement and system integration.

2. Overview of the Published Articles

Our ambient surroundings contain abundant mechanical energy sources that are not fully used in most circumstances, including vibrations, rotation, wind, water waves, raindrops and human motions. To efficiently harvest the mechanical energy in low-frequency ranges, Li et al. (contribution 1) proposed a piezoelectric rotational energy harvester with a dual-frequency up-conversion design. The energy harvester consists of multiple piezoelectric beams that undergo high-frequency self-excitations for efficient power generation under the trigger of a multi-paddle slider. The testing results indicate that the designed rotational energy harvester with three piezoelectric beams can produce an output power and a power density of 5.392 mW and 4.02 μW/(cm^3 Hz), respectively, when driven by a 0.42 Hz rotational excitation. Meanwhile, Zhang et al. (contribution 2) presented a push–pull diamagnetic levitation design for airflow energy harvesting using the electromagnetic mechanism. To compensate for the rotor offset and ensure higher efficiency, the floating rotor is configured with a symmetrical four-notch structure. The energy harvester is able to generate a peak voltage of 2.709 V (40.80% improvement over a three-notch structure) when subjected to a 3000 sccm airflow. The corresponding output power is 138.47 mW, showing great potential in scavenging the ambient airflow energy.

Radio frequency (RF) electromagnetic waves are another common form of renewable energy in the environment due to the widely deployed wireless communication networks. Li et al. (contribution 3) proposed a fully on-chip solution for tunable voltage boosting (TVB) to improve the acquisition performance of low-power RF signals. With the optimal TVB design, the output voltage of the rectifier can be maintained at a high level of 1 V across the new broad 5G radio frequency bandwidth from 3 to 6 GHz. Typically, an output voltage and a peak power conversion of 1 V DC and 83% are achieved at 3 GHz when an input power of -23 dBm is applied. Concurrently, the same group (contribution 4) further developed a low-dropout regulator (LDO) across multi-voltage domains using the 180 nm complementary metal oxide semiconductor (CMOS) technique for RF energy harvesting. The as-fabricated chip exhibits a small area of only 0.149 mm^2, although it is able to deliver a large current of up to 400 mA. For the low-voltage domain, the LDO has a line regulation of 1.85 mV/V and a load regulation of 0.0003 mV/mA. However, for the high-voltage domain, these parameters increase to 3.53 mV/V and 0.079 mV/mA. The small form factor and the low standby power consumption (174.5 μW) make the LDO an ideal candidate for powering IoT nodes using an RF energy harvesting chip.

To enable self-sustained IoT remote traffic sensing, Dipon et al. (contribution 5) presented a hybrid energy harvester that can simultaneously scavenge the mechanical vibration from passing vehicles and the temperature gradient from the asphalt and the soil underneath the road. The hybrid device consists of a stacked piezoelectric generator and a thermoelectric generator with respective AC-DC and DC-DC signal converters to regulate the generated outputs for battery recharging. The harvested energy from the dual sources in the ambient surroundings is sufficient to drive the operation of a multi-sensing system with wireless communication capability. A maximum communication range of 1 mile is obtained between the IoT node and the gateway in both laboratory tests and actual road conditions, indicating the high efficiency and good performance of the hybrid energy harvester in practical usage scenarios.

Due to the direct and converse piezoelectric effect, piezoelectric devices can be adopted for both self-powered sensors and actuators in more diversified applications. Wang et al. (contribution 6) reported a piezoelectric microelectromechanical system (MEMS) speaker consisting of a main functional lead zirconate titanate (PZT) layer and a designed supporting layer based on rigid–flexible coupling. Compared to its rigid counterpart, the rigid–flexible coupling layer provides the speaker with a higher sound pressure level (SPL) at low frequencies, e.g., a 4.1–20.1 dB improvement for 20 Hz to 4.2 kHz. In addition, the speaker can also serve as a silent alarm in dangerous situations via the detection of oral airflows, as it is able to recognize different words based on their induced signals with unique features, and it is highly resistant to ambient interference.

The use of a cognitive radio (CR) composed of a transceiver for detecting and moving into available communication channels is a common strategy in wireless communication. Devaraj et al. (contribution 7) presented a cluster-based cooperative spectrum sensing (CBCSS) advanced algorithm with detailed investigations to improve the achievable throughput in CR. The proposed CBCSS algorithm is demonstrated for 5G and beyond-5G IoT networks, showing a better performance in achievable throughput, average cluster number and energy than the currently adopted algorithms.

In addition to the use of energy harvesters for sustainable output generation, power management circuits from energy generation to storage are also indispensable for the realization of high-efficiency self-powered IoT systems. Zhang et al. (contribution 8) proposed the design and simulation of an LDO with an outstanding low dropout voltage (100 mV) and quiescent current (nA level). To ensure high reliability and transient response, adaptive power transistors and an adaptive bias are included in the design, and the simulation results show good performance and broad adaptability for energy harvesting systems. Then, Wu et al. (contribution 9) presented a boost converter circuit with only one single inductor for managing the outputs from thermoelectric energy harvesters. A two-stage startup circuit with a three-phase operation is adopted to obtain self-startup with the single inductor. A maximum power point tracking (MPPT) strategy with coarse and fine tuning is developed to achieve the most optimal energy extraction. The boost converter is then simulated in a 180 nm CMOS process, and the results indicate that it can achieve peak efficiency and a maximum power of 88% and 3.78 mW, respectively, after self-starting at an input voltage of 128 mV.

For those who are seeking a broader overview and a more general understanding of the power management strategies in energy harvesting, the review article by Lian et al. (contribution 10) could be a good reference. The authors discuss the effective methods used to extract energy from different energy harvesters, such as photovoltaic cells, thermoelectric generators, piezoelectric generators and RF energy harvesters. Based on the distinct characteristics of each type of energy harvester, actual implementation considerations and applicable circuit designs are systematically summarized, together with their pros and cons. In particular, MPPT as an effective strategy for system efficiency improvement is highlighted and compared with other circuits in detail, echoing the work conducted by Wu et al. Overall, the two promising directions for micro-energy harvesting are further improving management efficiency via low-loss accurate control and reducing the circuit size using capacitor-based solutions.

3. Conclusions

The articles published in this Special Issue present important advancements in energy harvesting and self-powered sensing technologies, and the article compilation contributes to broadening the domain knowledge in the field. Although numerous exciting achievements have been made, there are still some important aspects that need to be taken into account when looking at future directions. First, in addition to the further performance improvement of energy harvesting devices, their adaptability is also essential, as the implementation environment can be complicated and unpredictable. For example, devices designed with a single transducing mechanism may suffer from the fluctuation and intermittentness of the energy source, resulting in insufficient output generation and functionality failure. One feasible approach to enhance adaptability and mitigate this issue is a hybrid generator design that combines two or more transducing mechanisms for harvesting energy from multiple sources [11]. Second, the robustness and long-term reliability of devices should be considered in practical usage scenarios in order to enable their survival when applied in extreme environments, such as high-temperature, high-pressure and underwater environments. Careful material selection, structure design and proper encapsulation should be performed prior to practical deployment [12]. Last but not least, improving the intelligence of sensors and systems is necessary and a key enabler for various smart applications [13–15]. The thriving of artificial intelligence (AI) technology offers a good opportunity to integrate machine learning methodologies with IoT sensors and systems. The fusion of IoT and AI technology leads to the emergence and flourishing of artificial intelligence of things (AIoT), which will no doubt conversely push the development of IoT forward.

As a final note, we hope that the articles showcased in this Special Issue are interesting and inspiring for readers who want to learn about the recent progress and conduct new research in the field. With further advancements addressing the above aspects, we also hope to see the commercialization of more diverse energy-harvesting devices and self-powered IoT nodes in the near future. The widespread use of such devices will help further promote the large-scale deployment of IoT and AIoT systems and networks, ensuring the overall welfare of human beings and the digital transformation of industries and our society.

Funding: This work was funded by the National Natural Science Foundation of China (62301150), the Start-up Research Fund of Southeast University (RF1028623164) and the Nanjing Science and Technology Innovation Project for Returned Overseas Talent (4206002302).

Acknowledgments: We would like to express our sincere gratitude to all authors for contributing their excellent works to this Special Issue. In addition, we also want to thank all reviewers for their valuable dedication in the peer-review process, helping to maintain the high standard of the publications. Last but not least, we appreciate the great supportive assistance from the editorial team during this memorable period.

Conflicts of Interest: The authors declare no conflicts of interest.

List of Contributions:

1. Li, N.; Xia, H.; Yang, C.; Luo, T.; Qin, L. Investigation of a Novel Ultra-Low-Frequency Rotational Energy Harvester Based on a Double-Frequency Up-Conversion Mechanism. *Micromachines* **2023**, *14*, 1645. https://doi.org/10.3390/mi14081645.
2. Zhang, L.; Shao, H.; Zhang, J.; Liu, D.; Aw, K.C.; Su, Y. Improvement of the Airflow Energy Harvester Based on the New Diamagnetic Levitation Structure. *Micromachines* **2023**, *14*, 1374. https://doi.org/10.3390/mi14071374.
3. Li, Y.; Rajendran, J.; Mariappan, S.; Rawat, A.S.; Hamid, S.S.; Kumar, N.; Othman, M.; Nathan, A. CMOS Radio Frequency Energy Harvester (RFEH) with Fully On-Chip Tunable Voltage-Booster for Wideband Sensitivity Enhancement. *Micromachines* **2023**, *14*, 392. https://doi.org/10.3390/mi14020392.
4. Poongan, B.; Rajendran, J.; Li, Y.; Mariappan, S.; Parameswaran, P.; Kumar, N.; Othman, M.; Nathan, A. A 53-μA-Quiescent 400-mA Load Demultiplexer Based CMOS Multi-Voltage Domain Low Dropout Regulator for RF Energy Harvester. *Micromachines* **2023**, *14*, 379. https://doi.org/10.3390/mi14020379.
5. Dipon, W.; Gamboa, B.; Estrada, M.; Flynn, W.P.; Guo, R.; Bhalla, A. Self-Sustainable IoT-Based Remote Sensing Powered by Energy Harvesting Using Stacked Piezoelectric Transducer and Thermoelectric Generator. *Micromachines* **2023**, *14*, 1428. https://doi.org/10.3390/mi14071428.
6. Wang, Q.; Ruan, T.; Xu, Q.; Hu, Z.; Yang, B.; You, M.; Lin, Z.; Liu, J. A Piezoelectric MEMS Speaker with a Combined Function of a Silent Alarm. *Micromachines* **2023**, *14*, 702. https://doi.org/10.3390/mi14030702.
7. Devaraj, S.A.; Gurumoorthy, K.B.; Kumar, P.; Jacob, W.S.; Prince Jenifer Darling Rosita, P.J.D.; Ali, T. Cluster-ID-Based Throughput Improvement in Cognitive Radio Networks for 5G and Beyond-5G IoT Applications. *Micromachines* **2022**, *13*, 1414. https://doi.org/10.3390/mi13091414.
8. Zhang, Y.; Ge, Q.; Zeng, Y. A 0.6 V_{IN} 100 mV Dropout Capacitor-Less LDO with 220 nA I_Q for Energy Harvesting System. *Micromachines* **2023**, *14*, 998. https://doi.org/10.3390/mi14050998.
9. Wu, C.; Zhang, J.; Zhang, Y.; Zeng, Y. A 7.5-mV Input and 88%-Efficiency Single-Inductor Boost Converter with Self-Startup and MPPT for Thermoelectric Energy Harvesting. *Micromachines* **2023**, *14*, 60. https://doi.org/10.3390/mi14010060.
10. Lian, Q.; Han, P.; Mei, N. A Review of Converter Circuits for Ambient Micro Energy Harvesting. *Micromachines* **2022**, *13*, 2222. https://doi.org/10.3390/mi13122222.

References

1. Shafique, K.; Khawaja, B.A.; Sabir, F.; Qazi, S.; Mustaqim, M. Internet of Things (IoT) for Next-Generation Smart Systems: A Review of Current Challenges, Future Trends and Prospects for Emerging 5G-IoT Scenarios. *IEEE Access* **2020**, *8*, 23022–23040. [CrossRef]
2. Bai, Y.; Jantunen, H.; Juuti, J. Energy Harvesting Research: The Road from Single Source to Multisource. *Adv. Mater.* **2018**, *30*, 1707271. [CrossRef] [PubMed]
3. Liu, H.; Zhong, J.; Lee, C.; Lee, S.-W.; Lin, L. A Comprehensive Review on Piezoelectric Energy Harvesting Technology: Materials, Mechanisms, and Applications. *Appl. Phys. Rev.* **2018**, *5*, 041306. [CrossRef]
4. Wu, C.; Wang, A.C.; Ding, W.; Guo, H.; Wang, Z.L. Triboelectric Nanogenerator: A Foundation of the Energy for the New Era. *Adv. Energy Mater.* **2019**, *9*, 1802906. [CrossRef]

5. Dan, X.; Cao, R.; Cao, X.; Wang, Y.; Xiong, Y.; Han, J.; Luo, L.; Yang, J.; Xu, N.; Sun, J.; et al. Whirligig-Inspired Hybrid Nanogenerator for Multi-Strategy Energy Harvesting. *Adv. Fiber Mater.* **2023**, *5*, 362–376. [CrossRef]

6. Nozariasbmarz, A.; Collins, H.; Dsouza, K.; Polash, M.H.; Hosseini, M.; Hyland, M.; Liu, J.; Malhotra, A.; Ortiz, F.M.; Mohaddes, F.; et al. Review of Wearable Thermoelectric Energy Harvesting: From Body Temperature to Electronic Systems. *Appl. Energy* **2020**, *258*, 114069. [CrossRef]

7. Li, Q.; Li, S.; Pisignano, D.; Persano, L.; Yang, Y.; Su, Y. On the Evaluation of Output Voltages for Quantifying the Performance of Pyroelectric Energy Harvesters. *Nano Energy* **2021**, *86*, 106045. [CrossRef]

8. Roldán-Carmona, C.; Malinkiewicz, O.; Soriano, A.; Mínguez Espallargas, G.; Garcia, A.; Reinecke, P.; Kroyer, T.; Dar, M.I.; Nazeeruddin, M.K.; Bolink, H.J. Flexible High Efficiency Perovskite Solar Cells. *Energy Environ. Sci.* **2014**, *7*, 994. [CrossRef]

9. Zhang, X.; Grajal, J.; Vazquez-Roy, J.L.; Radhakrishna, U.; Wang, X.; Chern, W.; Zhou, L.; Lin, Y.; Shen, P.C.; Ji, X.; et al. Two-Dimensional MoS_2-Enabled Flexible Rectenna for Wi-Fi-Band Wireless Energy Harvesting. *Nature* **2019**, *566*, 368–372. [CrossRef] [PubMed]

10. Zhu, M.; Yi, Z.; Yang, B.; Lee, C. Making Use of Nanoenergy from Human–Nanogenerator and Self-Powered Sensor Enabled Sustainable Wireless IoT Sensory Systems. *Nano Today* **2021**, *36*, 101016. [CrossRef]

11. Shi, Q.; Sun, Z.; Zhang, Z.; Lee, C. Triboelectric Nanogenerators and Hybridized Systems for Enabling Next-Generation IoT Applications. *Research* **2021**, *2021*, 6849171. [CrossRef] [PubMed]

12. Zhao, H.; Xu, M.; Shu, M.; An, J.; Ding, W.; Liu, X.; Wang, S.; Zhao, C.; Yu, H.; Wang, H.; et al. Underwater Wireless Communication via TENG-Generated Maxwell's Displacement Current. *Nat. Commun.* **2022**, *13*, 3325. [CrossRef] [PubMed]

13. Jin, T.; Sun, Z.; Li, L.; Zhang, Q.; Zhu, M.; Zhang, Z.; Yuan, G.; Chen, T.; Tian, Y.; Hou, X.; et al. Triboelectric Nanogenerator Sensors for Soft Robotics Aiming at Digital Twin Applications. *Nat. Commun.* **2020**, *11*, 5381. [CrossRef] [PubMed]

14. Askari, H.; Khajepour, A.; Khamesee, M.B.; Wang, Z.L. Embedded Self-Powered Sensing Systems for Smart Vehicles and Intelligent Transportation. *Nano Energy* **2019**, *66*, 104103. [CrossRef]

15. Liu, L.; Guo, X.; Lee, C. Promoting Smart Cities into the 5G Era with Multi-Field Internet of Things (IoT) Applications Powered with Advanced Mechanical Energy Harvesters. *Nano Energy* **2021**, *88*, 106304. [CrossRef]

 *micromachines*

MDPI

Article

Investigation of a Novel Ultra-Low-Frequency Rotational Energy Harvester Based on a Double-Frequency Up-Conversion Mechanism

Ning Li [1,2], Hu Xia [1,2], Chun Yang [1,2], Tao Luo [1,2] and Lifeng Qin [1,2,*]

[1]   Shenzhen Research Institute of Xiamen University, Shenzhen 518000, China;
     19920221151543@stu.xmu.edu.cn (N.L.); frighten@stu.xmu.edu.cn (H.X.);
     199202111151546@stu.xmu.edu.cn (C.Y.); luotao@xmu.edu.cn (T.L.)
[2]   Department of Mechanical and Electrical Engineering, Xiamen University, Xiamen 361005, China
*   Correspondence: liq@xmu.edu.cn

Abstract: Due to their lack of pollution and long replacement cycles, piezoelectric energy harvesters have gained increasing attention as emerging power generation devices. However, achieving effective energy harvesting in ultra-low-frequency (<1 Hz) rotational environments remains a challenge. Therefore, a novel rotational energy harvester (REH) with a double-frequency up-conversion mechanism was proposed in this study. It consisted of a hollow cylindrical shell with multiple piezoelectric beams and a ring-shaped slider with multiple paddles. During operation, the relative rotation between the slider and the shell induced the paddles on the slider to strike the piezoelectric beams inside the shell, thereby causing the piezoelectric beams to undergo self-excited oscillation and converting mechanical energy into electrical energy through the piezoelectric effect. Additionally, by adjusting the number of paddles and piezoelectric beams, the frequency of the piezoelectric beam struck by the paddles within one rotation cycle could be increased, further enhancing the output performance of the REH. To validate the output performance of the proposed REH, a prototype was fabricated, and the relationship between the device's output performance and parameters such as the number of paddles, system rotation speed, and device installation eccentricity was studied. The results showed that the designed REH achieved a single piezoelectric beam output power of up to 2.268 mW, while the REH with three piezoelectric beams reached an output power of 5.392 mW, with a high power density of 4.02 μW/(cm³ Hz) under a rotational excitation of 0.42 Hz, demonstrating excellent energy-harvesting characteristics.

Keywords: piezoelectric; energy harvesting; rotational; ultra-low frequency

Citation: Li, N.; Xia, H.; Yang, C.; Luo, T.; Qin, L. Investigation of a Novel Ultra-Low-Frequency Rotational Energy Harvester Based on a Double-Frequency Up-Conversion Mechanism. *Micromachines* **2023**, *14*, 1645. https://doi.org/10.3390/mi14081645

Academic Editors: Qiongfeng Shi and Huicong Liu

Received: 12 July 2023
Revised: 18 August 2023
Accepted: 19 August 2023
Published: 20 August 2023

1. Introduction

With the widespread adoption of 5G networks and the rapid development of Internet of Things (IoT) technologies, numerous electronic devices have been applied in various corners of the world. However, the increasing demand for power supply among these electronic devices poses a challenging problem. Traditional battery-powered solutions for electronic devices working in complex and harsh environments, such as various sensors, suffer from issues like short lifespans, high replacement costs, and environmental pollution after disposal. Therefore, there is an urgent need for a new type of power device that can provide a long-term energy supply. Energy harvesters, as emerging energy supply devices, leverage various mechanisms, such as piezoelectric [1–3], electromagnetic [4,5], triboelectric [6,7], and photovoltaic [8,9] effects, to convert mechanical energy, wind energy, solar energy, and other forms of ambient energy into electrical energy for external output. Among them, piezoelectric energy harvesters [10–12] have been employed in widespread applications due to their simple structure [13,14], long lifespan [15,16], and high adaptability [17–19] to environmental conditions. These devices operate by utilizing the mechanical energy obtained from the external environment to induce deformation in the internal piezoelectric material. Subsequently, the piezoelectric effect of the material is harnessed to convert the collected mechanical energy into electrical energy and output it externally. Depending on the type of motion in the application environment, such as vibration [20,21], oscillation [22,23], and rotation [24,25], piezoelectric energy harvesters can be classified as vibration energy harvesters [26,27] and rotational energy harvesters [28–30]. Extensive progress has been made in the research on vibration-based energy harvesters. Meanwhile, although it lags behind research into vibration-based energy harvesters, in order to meet the growing demand for energy conversion and utilization, research in the field of rotation-based energy harvesters is also rapidly advancing.

In the field of piezoelectric rotational energy harvesters (REHs), three types can be identified based on the mode of forced deformation of the piezoelectric material: magnetic-driven [31,32], gravity-driven [33,34], and mechanical-plucking-driven [35–37] REHs. For instance, in the work by Zou et al. [38], researchers employed a disc rotating around the center with magnets placed on its circumference as the rotor, while a bridge-like structure containing piezoelectric material and magnets was fixed around the disc as the stator. During the operation of the REH, as the magnets on the rotor approached the magnets on the stator, the bridge-like structure deformed under magnetic forces, thereby inducing compression and tension on the internal piezoelectric material for energy conversion. However, the use of magnetic-driven REHs is limited when the application environment is sensitive to magnetic fields. Conversely, REHs that utilize gravity for external excitation to generate the output can effectively avoid this issue. In the work by Guan et al. [39], researchers fixed one end of the piezoelectric beam on the rotor, while another end had a mass block attached, with the center of gravity aligned with the rotation center. During the rotation of the REH, the periodic deformation of the piezoelectric beam was driven by the gravitational force, converting mechanical energy into electrical energy. The results demonstrated that the REH achieved a power density of 0.81 μW/(cm^3 Hz) at an excitation frequency of 0.79 Hz. However, due to the gravitational excitation period matching the system's rotational period, the output performance of the REH was limited in ultra-low-frequency rotational environments. In contrast, the mechanical-plucking approach often allows the piezoelectric beam to undergo self-excited oscillation at its inherent frequency after being mechanically plucked, thereby enhancing the output capability of the REH in low-frequency rotational environments. In the work by Fang et al. [40], researchers designed a rotating energy harvester composed of a cylindrical rotor with multiple plucking elements and a stator with multiple piezoelectric beams. Similar to the mechanism of a music box, when the REH was in operation, the motion of the rotor at the rotation center triggered the plucking elements to strike the piezoelectric beams, inducing self-excited oscillation for energy conversion. However, this fixed-rotor or stator arrangement necessitated mounting one of them at the rotation center. In certain rotational environments where the installation of external mechanisms is impractical due to structural constraints (e.g., car wheels) or where the devices requiring powering are located far from the rotation center, the application of such REHs is limited. To address such issues, in our previous work [41], we designed and fabricated an REH that utilized liquid as the energy-capturing medium. The proposed REH included a piezoelectric beam with a flow resistance plate, a cylindrical shell, and a liquid. The fluid's flow characteristics enabled it to impact the piezoelectric beam at low rotational speeds and act as a mass block at high rotational speeds, inducing the periodic deformation of the internal piezoelectric material and generating an electric output based on the piezoelectric effect. The results demonstrated that the REH achieved a power density of 0.23 μW/(cm^3 Hz) at an excitation frequency of 1.25 Hz. However, this was still insufficient for effectively powering devices with higher energy demands. In conclusion, although significant progress has been made in the research on rotational energy harvesters, achieving higher power outputs in low-frequency rotational environments (<1 Hz), such as wind turbines and hydro turbines [13,41], remains a challenging issue to be addressed.

In this study, we proposed a novel ultra-low-frequency rotational energy harvester based on a double-frequency up-conversion mechanism. The harvester consisted of a cylindrical shell internally equipped with multiple piezoelectric beams and a slider with multiple paddles attached. During the movement of the rotational energy harvester (REH) along with the rotating system, a relative rotation occurred between the slider and the shell due to the influence of gravity, resulting in the activation of the paddles to strike the piezoelectric beams, inducing the self-excited oscillation of the beams (the first frequency up-conversion) and enabling the conversion of mechanical energy into electrical energy through the utilization of the piezoelectric effect. Simultaneously, by appropriately increasing the number of paddles on the slider and the number of internal piezoelectric beams in the REH, a significant enhancement in the collision frequency between the paddles and piezoelectric beams within one rotation cycle could be achieved (the second frequency up-conversion), thereby further improving the output performance of the REH. During the experiment, the proposed REH achieved an external output power of 5.392 mW and a power density of 4.02 μW/(cm^3 Hz) under a rotational excitation of 0.42 Hz. Additionally, the proposed design of the REH allowed for the retention of the relative motion between the internal slider and shell even when the REH was eccentrically installed away from the rotational center of the system. Therefore, under the conditions of eccentric installation, the REH remained capable of harvesting rotational energy. The paper is organized as follows: Section 2 discusses the working principle of the proposed energy harvester. Section 3 presents the fabrication process and experimental setup. Section 4 provides a detailed analysis of the results and includes a comprehensive discussion. Finally, Section 5 presents the meaningful conclusions drawn from the findings.

2. Working Principle of Proposed REH

As shown in Figure 1a, the proposed REH consisted of a cylindrical shell and a slider with paddles. Inside the cylindrical shell, there was a clamping platform designed to hold the piezoelectric beams. As shown in Figure 1b, the proposed REH was installed in the rotational system and rotated together with the system. Figure 1c illustrates the relative motion between the internal slider and the shell of the REH during this process. Simultaneously, we captured the open-circuit voltage waveform output by the REH within one rotation cycle using an oscilloscope, as shown in Figure 1d. Combining Figure 1c,d, it can be observed that the complete output process of the device in each rotation cycle consisted of three stages. The first stage occurred when the slider approached the piezoelectric beams, at which point the piezoelectric beams had no external output. The second stage was the contact between the paddles and the piezoelectric beams, where the piezoelectric beams rapidly reached their maximum deformation under the pushing force exerted by the paddles on the slider. Subsequently, the process entered the third stage, where the paddles detached from the piezoelectric beams, and the piezoelectric beams underwent self-excited oscillation with their inherent frequency (the first frequency up-conversion). It was in this stage that the mechanical energy was converted into electrical energy by the piezoelectric effect [42]. However, in an ultra-low-frequency rotational environment (<1 Hz), there is a long idle period between collisions due to the longer rotation period, during which the piezoelectric beam remains idle and does not generate an output. This phenomenon was also confirmed by the waveform of the output voltage obtained from the experiments, as shown in Figure 1d, which demonstrated a long interval between adjacent waveforms, indicating the significant proportion of the collision process taken up by the idle stage. This was caused by the longer rotation period at low frequencies and had a negative impact on the output performance of the REH.

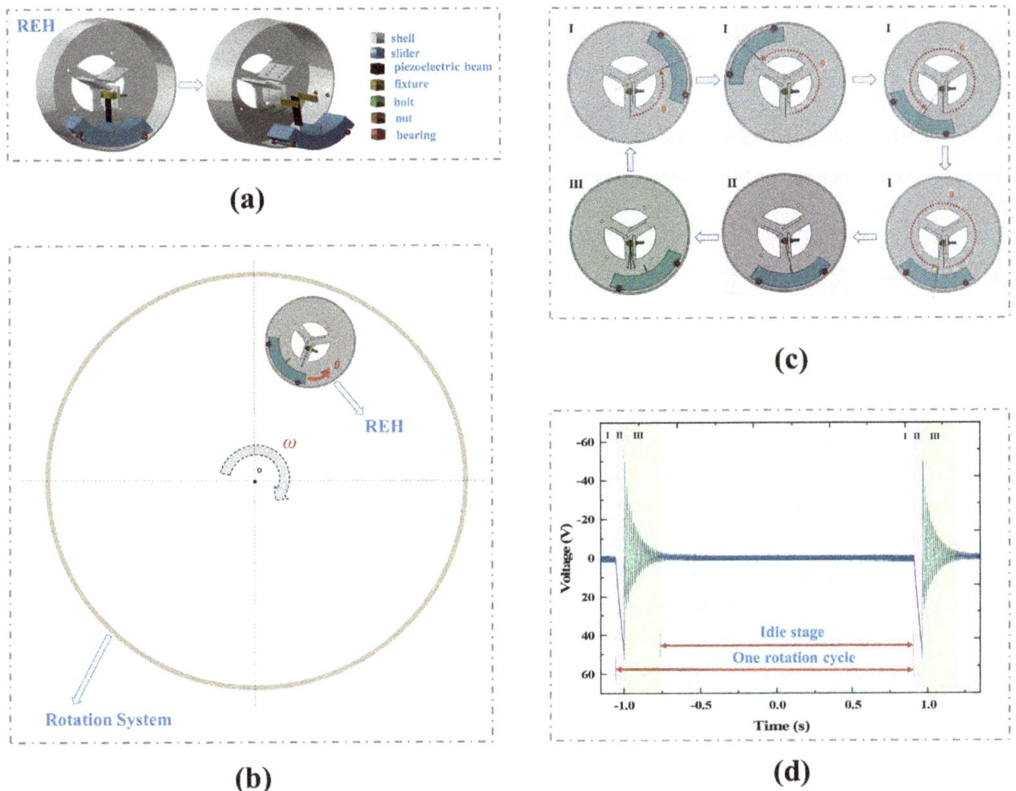

Figure 1. Operating principle of the REH: (**a**) composition of the REH; (**b**) REH in the rotating system; (**c**) operating principle of the REH; (**d**) open−circuit voltage waveform of the REH (system rotation speed ω = 30 rpm, number of paddles n = 1, number of piezoelectric beams k = 1).

Therefore, in order to enhance the output performance of the proposed REH, our goal was to minimize the proportion of the "idle stage" during the operation of the REH. We proposed a feasible solution to achieve a reduction in the "idle stage" by increasing the number of paddles attached to the slider, thereby increasing the frequency of the piezoelectric beams being actuated within one rotation cycle. Additionally, by increasing the number of piezoelectric beams inside the REH, the number of times the beams were actuated within one rotation cycle could also be increased. Therefore, when the REH contained k piezo beams and n paddles, the number of external outputs per rotation cycle reached $k \times n$ (the second frequency up-conversion), as shown in Figure 2a (for example, when n = 5 and k = 3, the REH obtained is shown in Figure 2b).

In order to validate the feasibility of the design concept, we conducted preliminary experiments using an REH equipped with a single piezoelectric beam. The REH-4 (an REH with four attached paddles, n = 4) was employed for testing, and the obtained waveforms are depicted in Figure 3a. Under the same rotational speed (30 rpm), as the number of paddles increased, the time interval between adjacent waveforms decreased. This indicated a higher proportion of effective working time for the REH within an ultra-low-frequency rotational environment. Therefore, increasing the number of paddles was a feasible approach to enhance the output performance of the proposed REH. Additionally, we tested the output waveforms of the device with different numbers of paddles and higher system rotational speeds, as shown in Figure 3b,c. From the figures, it can be observed that interference occurred between adjacent waveforms when the number of paddles or rotational speed exceeded a certain threshold. The impact of this interference on the REH's performance was further investigated in subsequent experiments.

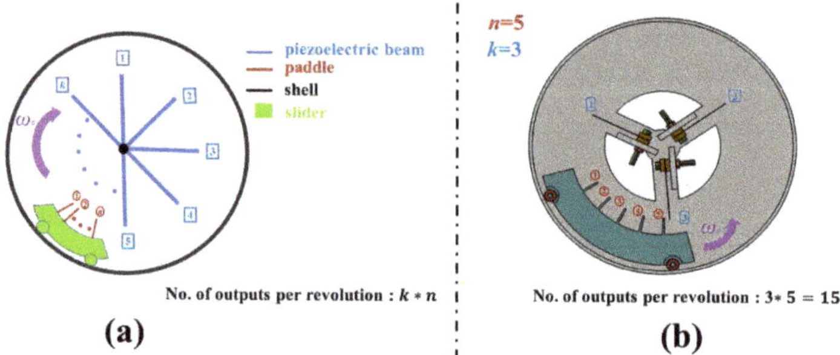

Figure 2. Prototype of the proposed REH (the numbers in the red labels correspond to the paddle indices, while the numbers in the blue labels correspond to the piezoelectric beam indices): (**a**) proposed energy harvester concept; (**b**) schematic drawing of the proposed REH (number of piezoelectric beams $k = 3$, number of paddles $n = 5$).

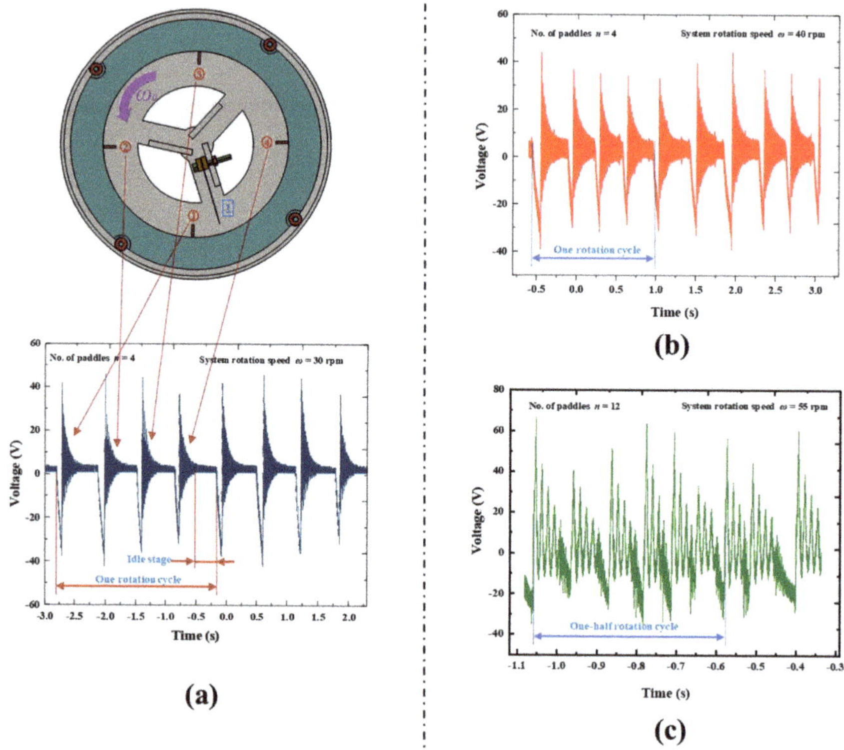

Figure 3. Open−circuit voltage waveforms of the REH at various rotational speeds and numbers of paddles (number of piezoelectric beams $k = 1$, the numbers in the red labels correspond to the paddle indices, while the numbers in the blue labels correspond to the piezoelectric beam indices.): (**a**) system rotation speed $\omega = 30$ rpm, number of paddles $n = 4$; (**b**) system rotation speed $\omega = 40$ rpm, number of paddles $n = 4$; (**c**) system rotation speed $\omega = 55$ rpm, number of paddles $n = 12$.

To further understand the working mechanism of the proposed REH, we needed to perform a dynamic analysis of its motion process. However, a similar dynamic model was established in our previous work [41], so we only provide a brief introduction to the establishment of the dynamic model in this paper.

Prior to the dynamic analysis, we simplified the model. Firstly, since the collision between the paddles and the piezoelectric beams was an instantaneous loading and unloading process, which depended on factors such as the mass of the slider and the motion state of the rotational system, it was a complex action process. Therefore, the influence of this process on the theoretical modeling was not considered temporarily in the dynamic analysis. Secondly, because the motion states of different parts of the slider were relatively consistent during its movement, and the shape of the slider had a minor impact on the motion process, the dynamic analysis could adopt a point mass to represent the slider, with the position of the point mass coinciding with the center of mass of the slider. On this basis, kinetic parameters were introduced into the system. As shown in Figure 4, the proposed REH was installed on a rotational system with angular velocity ω and an eccentricity distance of l_0 (the distance from the center of the rotation system (point O) to the center of the cylindrical shell (point A)). A relative coordinate system XAY was established inside the REH, and the relevant parameters are labeled accordingly. Meanwhile, specific definitions of the parameters are provided in Table 1.

Table 1. Relevant parameters in Figure 4.

Symbols	Meaning
ω	System angular velocity
t	System motion time
l_0	Installation eccentricity of the REH
l	Eccentricity of the slider's centroid
a	Acceleration of the slider relative to the shell
g	Gravitational acceleration
θ	Angular displacement of the centroid relative to the center of the shell
θ_1	Angular displacement of the centroid relative to the center of rotation
θ_2	$\theta - \theta_1$
F_N	Support force from shell
F_c	Centrifugal force of slider
m	Mass of slider
R	Radius of the cylindrical shell
r	Distance from the centroid of the slider to the bottom of the shell
F_k	Coriolis force of slider
F_f	Frictional force

Building upon this foundation, we employed the same modeling approach as in our previous work [41] to obtain the final results from the dynamic analysis.

In the relative coordinate system (XY) with the origin located at the center (point A) of the cylindrical shell and synchronized with the rotational system, the dynamic equation for the slider could be obtained through the analysis of the motion process as follows:

$$m\ddot{\theta}(R - r) = mg\sin(\omega t - \theta) - F_c\sin\theta_2 - \mu\left(m\omega^2 l\cos\theta_2 + mg\cos(\omega t - \theta) + 2m\omega v\right) \quad (1)$$

where μ is the friction coefficient. Meanwhile, the relationship between θ and θ_2 could be expressed as

$$\theta = \sin^{-1}\left(\frac{OA}{AB}\sin\theta_2\right) + \theta_2 \quad (2)$$

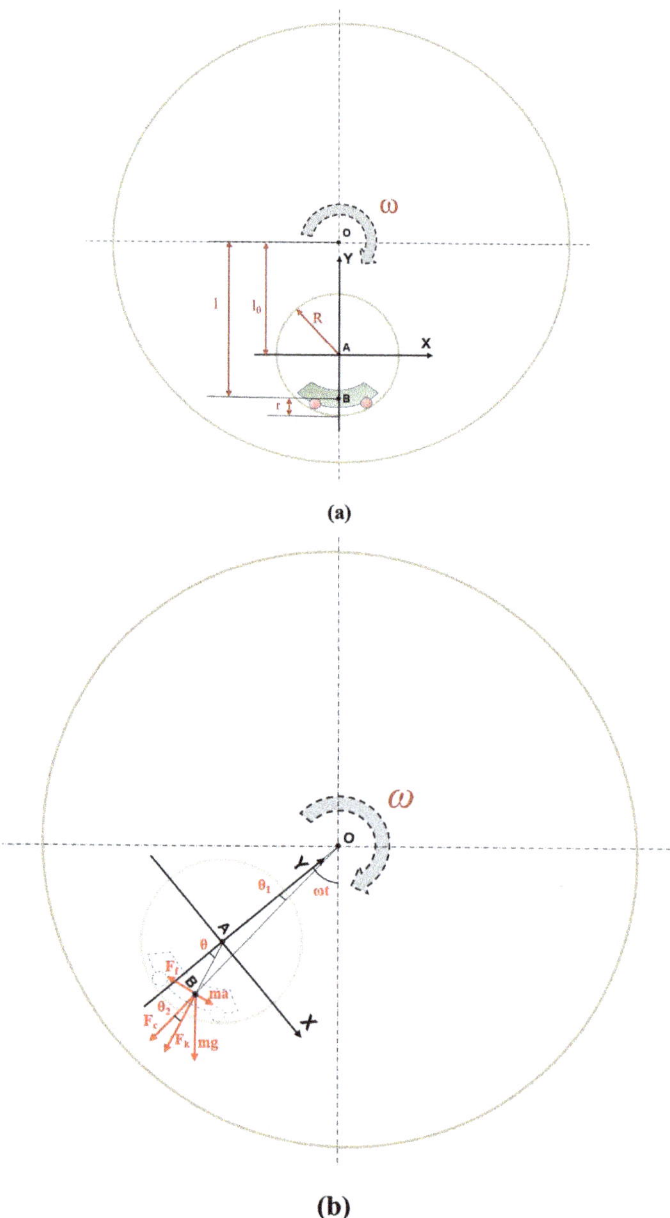

(a)

(b)

Figure 4. Dynamic model illustration of the REH operation process: (**a**) dimensional parameters of the system; (**b**) dynamic parameters of the system.

Based on Equations (1) and (2), the variation in the relative angular displacement between the center of mass of the slider and the shell with time could be obtained. Subsequently, the MATLAB Simulink module could be used to simulate the variation in the relative angular displacement with time. The simulation results are shown in Figure 5. Under low-speed conditions (taking the system speed as 30 rpm, $l_0 = 120$ mm, $R = 53$ mm, and $r = 20$ mm), the angular displacement θ of the slider increased with time, while the angular velocity $\dot{\theta}$ of the slider remained nearly constant during the rotation of the system. This phenomenon was also verified in practical experiments. Based on the analysis above, we could draw the conclusion that at lower system rotation speeds (<1 Hz), the relative rotational speed between the slider and the shell remained stable and numerically equal to the system's rotation speed. This result was also verified in subsequent experiments, as depicted by the waveforms in Figure 3.

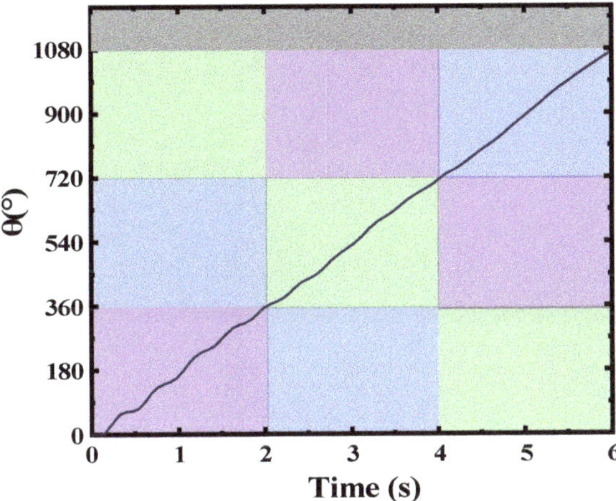

Figure 5. Displacement of the slider with time in angular motion.

3. Fabrication and Experimental Setup

In order to verify the feasibility of the proposed REH for effective energy harvesting in low-speed environments, a prototype was fabricated as shown in Figure 6b. It consisted of a hollow cylindrical shell and an internally hollow cylindrical slider, with the internal structure depicted in Figure 6a. Within the cylindrical shell, three holding platforms with an angular spacing of 120° were designed. Piezoelectric beams were installed on these platforms using fixtures and fasteners, allowing for adjustable displacement along the arrangement direction of the through holes on the platforms. As for the cylindrical slider, it featured four wheel shafts with an angular spacing of 90° on the outer side, along with four sets of bearings to reduce the frictional force between the slider and the shell. Simultaneously, the inner side of the slider contained 24 slots with an angular spacing of 15°, and six cavities with an angular spacing of 60° were designed within the slider using partition plates. The slots were utilized for mounting paddles. At the same time, the bottom of each paddle was designed in a cylindrical shape to match the array of through holes at the bottom of the slots, thereby adjusting the distance between the top of the paddle and the center of the shell. The number of paddles and the angular spacing between them could be adjusted by matching the slots with the paddles. Furthermore, the cavities were used to accommodate weights, ensuring that the mass center of the slider was positioned at the lower end, thus guaranteeing the stability of relative motion between the slider and the shell. The mass of the weights is represented as G.

The shell, slider, and paddles were all produced via 3D-printing technology, utilizing UV-curable resin material (SOMOS Imagine 8000, Royal DSM Group, Heerlen, the Netherlands). The cylindrical shell had a diameter of 231 mm, a height of 76 mm, and an inner-wall thickness of 2 mm. The piezoelectric cantilever beam (PZT 5J S118-J1SS-1808YB) in the device was produced by Mide Technology Company in the United States,, with detailed parameters provided in Table 2. The outer dimensions of the slider were a 217 mm diameter, 150 mm inner diameter, 65 mm axial thickness, and a wall thickness of 2 mm. The profile dimensions of the paddles were 65 mm × 50 mm × 2 mm.

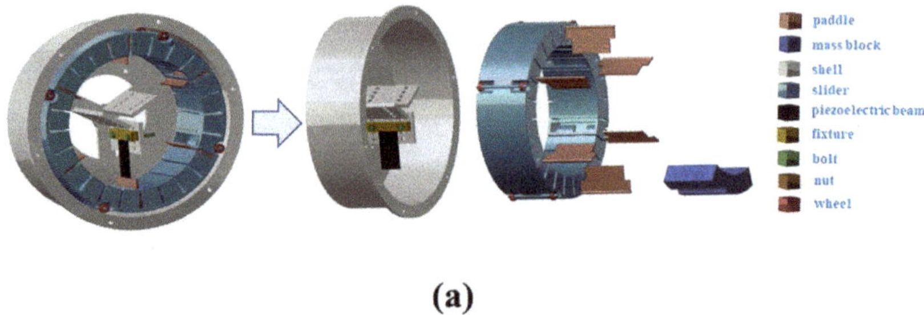

(a)

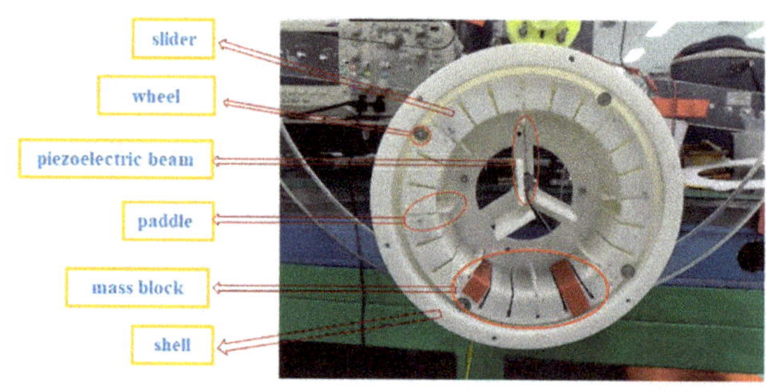

(b)

Figure 6. Prototype of the REH used in the experiments: (**a**) internal structure of the REH; (**b**) physical prototype of the proposed REH.

Table 2. Parameters of the beam structure and piezoelectric elements (PZT 5J S230-J1FR-1808XB, Made Technology Co., Woburn, MA, USA).

Parameter	Value
Length of beam	55.3 mm
Width of beam	23.3 mm
Thickness of beam	0.46 mm
Length of PZT 5J	46 mm
Width of PZT 5J	20.8 mm
Thickness of PZT 5J	0.15 mm
Resonant frequency	130 Hz
Spring constant	0.25 N/mm

To test the performance of the REH, a comprehensive testing system was established, as shown in Figure 7. The rotational excitation was provided by a servo motor, with the rotational speed controlled by a controller. The REH was installed on an acrylic disc with a diameter of 1200 mm, and the eccentricity distance of the REH was adjusted through an array of small holes on the disc. One REH and one counterweight were mounted on the left and right sides of the disc, respectively, and balanced using the lever principle to ensure the stable rotation of the entire system. The output (V_{RMS}) of the piezoelectric beams inside the REH was measured and recorded by connecting wires through a slip ring to an oscilloscope (Keysight DSO-X 2024A, Keysight Technologies, Santa Rosa, CA, USA).

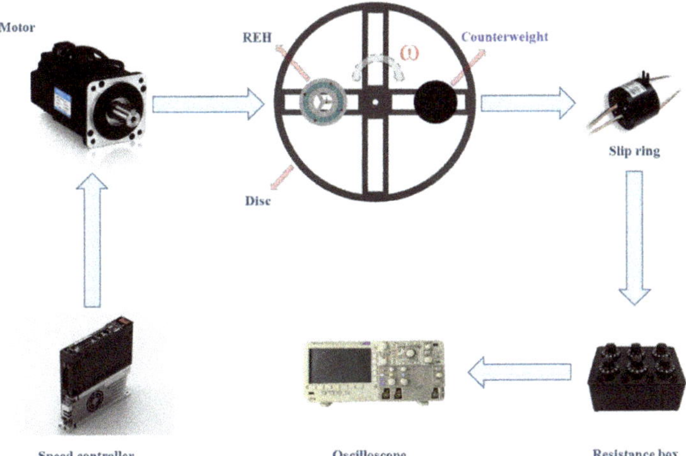

Figure 7. Experimental testing system for the REH.

4. Results and Discussion

To investigate the feasibility of enhancing the output performance of a rotational energy harvester (REH) in ultra-low-frequency rotating environments by increasing the number of paddles, we tested the output characteristics of an REH with a piezoelectric beam under the experimental conditions of a system speed of 25 rpm and eccentricity distance of 160 mm. The V_{RMS} (root mean square of voltage) results are shown in Figure 8. It can be observed that as the number of paddles increased from 1 to 8, the effective output voltage of the REH rapidly rose from 4.25 V to 11 V. However, as the number of paddles further increased, the rising trend weakened. The open-circuit voltages obtained for 12 and 24 paddles were 11.8 V and 12.4 V, respectively. These represent only a 7.3% and 12.7% increment compared to REH-8, despite having 1.5 and 3 times more paddles, respectively. This phenomenon could be attributed to the decreasing time interval for the piezoelectric beam to be actuated as the number of paddles increased. This was manifested in the output waveform shown in Figure 3c, where with an increasing number of paddles, an overlap between adjacent output waveforms became apparent. This led to energy loss, thereby limiting the improvement in the REH's output performance.

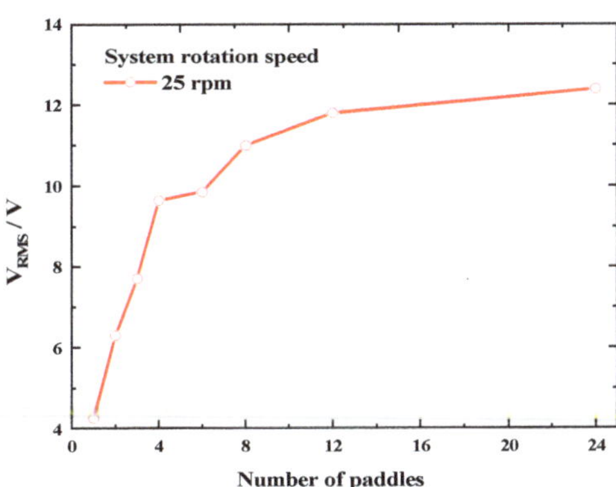

Figure 8. Variation in open-circuit voltage according to the number of paddles (system rotation speed ω = 25 rpm).

To further validate the relationship between the number of paddles and the output performance, we studied the variation in power with the number of paddles at the optimal load for the REH. First, under the experimental conditions of a system speed of 25 rpm and an eccentricity distance of 160 mm, we tested the optimal load characteristics of an REH with 2, 4, and 8 paddles. As shown in Figure 9, the optimal load (R) values for all three configurations were around 18 kΩ, indicating that the number of paddles did not alter the optimal load of the device.

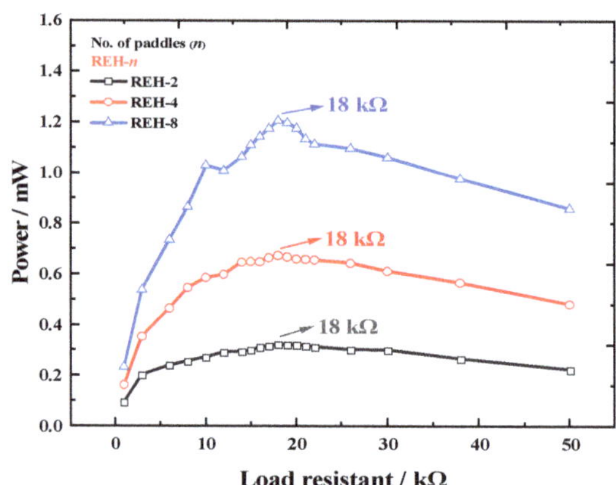

Figure 9. Output power of the REH as a function of load under different paddle numbers (system rotation speed ω = 25 rpm).

Based on this, we proceeded to test the output power of the REH with different numbers of paddles at the optimal load. Figure 10 illustrates that as the number of paddles increased from 1 to 8, the P_{RMS} (root mean square of power) of the REH at the optimal load rose rapidly from 0.153 mW to 1.206 mW, reaching 7.9 times the initial output value. However, as the number of paddles further increased, the rising trend weakened. The output powers obtained for 12 and 24 paddles were 1.313 mW and 1.514 mW, respectively, representing only an 8.9% and 25.5% increment compared to REH-8. This changing trend was similar to the voltage variation trend shown in Figure 8, and the underlying reason was the same, thus avoiding repetition.

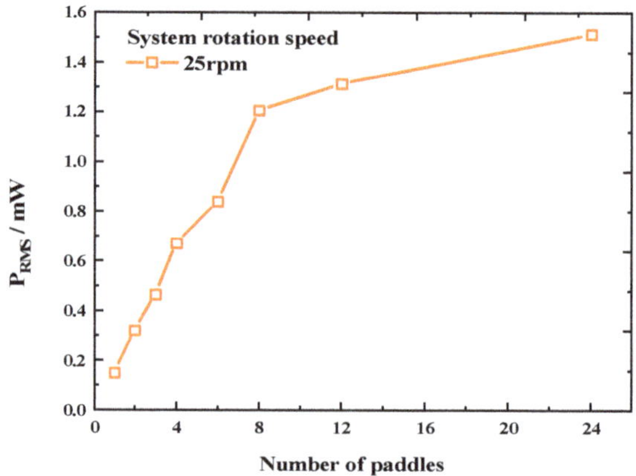

Figure 10. Variation in output power according to the number of paddles (system speed ω = 25 rpm).

From the aforementioned experimental results, it is evident that the number of paddles had a significant impact on the output performance of the REH in the range of 1 to 8. To further validate this output characteristic, we investigated the effects of parameters such as rotational speed and eccentricity distance on the REH output performance while varying the number of paddles.

To explore the influence of the rotational speed on the REH output performance, we initially studied the optimal load characteristics of the REH at different speeds. Taking REH-8 as the test specimen, we performed measurements of the optimal load under four different speeds (5 rpm, 15 rpm, 25 rpm, and 35 rpm) when the eccentricity distance was set at 160 mm. The results shown in Figure 11 indicate that the optimal load values for the device remained around 19 kΩ for 5 rpm and 15 rpm conditions, while for 25 rpm and 35 rpm conditions, the optimal load values were around 18 kΩ. Thus, at lower rotational speeds, the REH's optimal load remained between 18 kΩ and 19 kΩ.

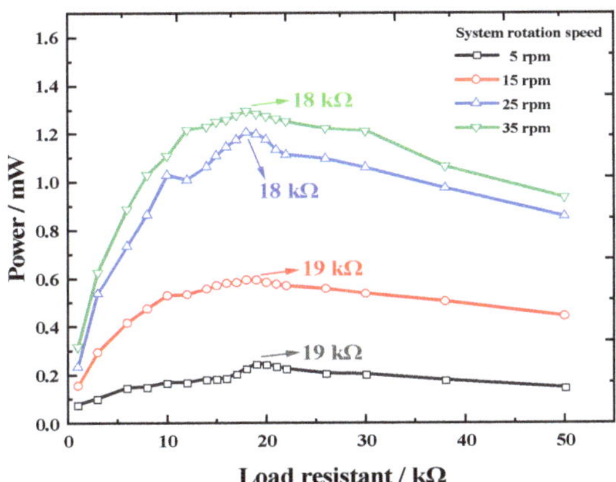

Figure 11. Output power of the REH as a function of load under different rotational speeds (number of paddles $n = 8$).

Based on this, we conducted tests on the optimal output of the device at different speeds. Figure 12 demonstrates that as the speed increased, the REH's output power at the optimal load continued to rise. As the speed increased from 5 rpm to 40 rpm, the REH's output power increased from 0.246 mW to 1.402 mW, representing a 5.7-fold increment. However, as the speed further increased, the upward trend in output power weakened. At a speed of 50 rpm, the output power of the device was 1.461 mW, which was only a 4.2% increase compared to the output at 40 rpm. The possible cause for this trend could have been that as the system rotation speed increased, the frequency at which the piezoelectric beams were struck also increased, leading to mutual interference between adjacent strikes. This ultimately resulted in a decrease in the rising trend of the device's output power, as demonstrated in the waveform shown in Figure 3c.

Additionally, we investigated the relationship between the output performance of the REH and the number of paddles under different rotation speeds and eccentricity conditions, and the experimental results are shown in Figures 13 and 14. As shown in Figure 13, at different rotation speeds, the output power of the REH under the optimal load increased with an increasing number of paddles. Among them, REH-8 achieved a maximum power output of 1.296 mW at a speed of 35 rpm. Similarly, as shown in Figure 14, under different eccentricity conditions, the device's output power also increased with an increasing number of paddles. Therefore, within a certain range of values, variations in rotation speed and eccentricity did not significantly affect the improvement in output performance due to the number of paddles.

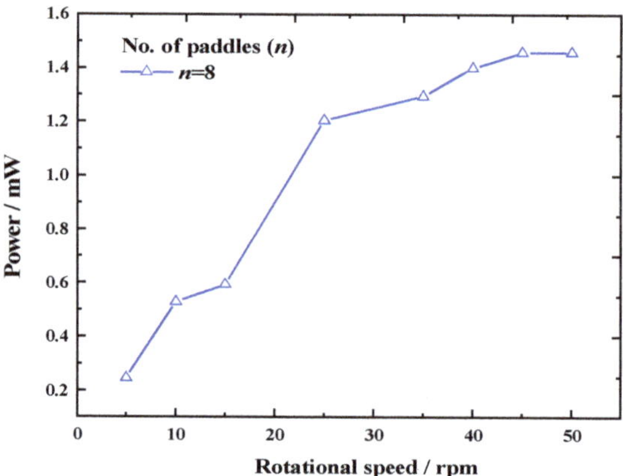

Figure 12. Variation in output power according to rotational speed (number of paddles $n = 8$).

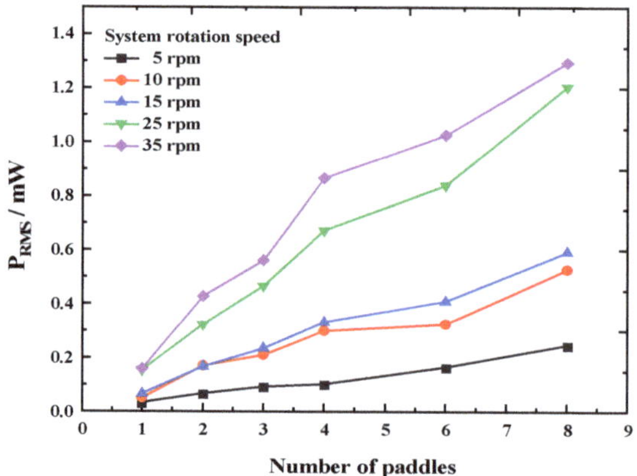

Figure 13. Output power of the REH as a function of the number of paddles under different rotational speeds.

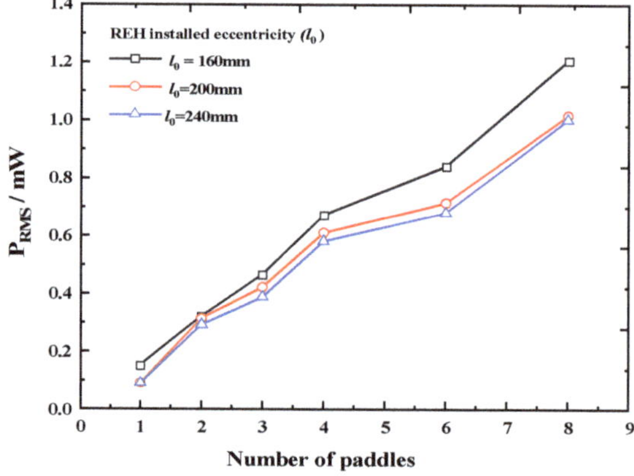

Figure 14. Output power of the REH as a function of the number of paddles under different eccentricities (system rotation speed $\omega = 25$ rpm).

Hence, we concluded that increasing the number of paddles was a feasible approach to enhancing the output performance of the REH. Furthermore, to further improve the output performance, we employed an array structure to increase the number of piezoelectric beams inside the REH. Figure 15a,b present schematic diagrams of the REH structure with $n = 2$ and $n = 3$, respectively. Simultaneously, to obtain the P_{RMS} of a multi-piezoelectric-beam REH, we sequentially measured the output voltage of each individual piezoelectric beam and calculated the output power (as shown in Figure 15c,d); then, we summed the power of each piezoelectric beam to obtain the total output power of the REH ($P_{RMS} = \sum_{x=1}^{k} \frac{V_{Rx}}{Rx}$, V_R represents the effective voltage across the optimal load (R)). However, frequent collisions could potentially result in a loss of kinetic energy for the slider, which might have undermined the effectiveness of individual impacts and subsequently impacted the output performance of the REH. As the kinetic energy of the slider was directly proportional to its mass when the velocity was constant, we investigated the impact of the mass block mass (G) on the output performance of the REH (system rotation speed $\omega = 30$ rpm, number of paddles $n = 8$, number of piezoelectric beams $k = 1$), as depicted in Figure 16. When $G \leq 100$ g, the paddles within the REH were unable to drive the piezoelectric beams into self-excited oscillations during its operational process. As a result, the slider and the shell remained relatively stationary during system rotation, leading to minimal external output from the REH. At $G = 200$ g, the paddles were able to induce self-excited oscillations in the piezoelectric beams, generating external output. However, as G was further increased (>200 g), the REH's output did not exhibit a growth trend ($V_{RMS} = (9.6 \pm 0.5)$ V). This indicates that once the mass block mass was sufficient to initiate self-excited oscillations in the piezoelectric beams, the impact of G on the REH's output became relatively minor.

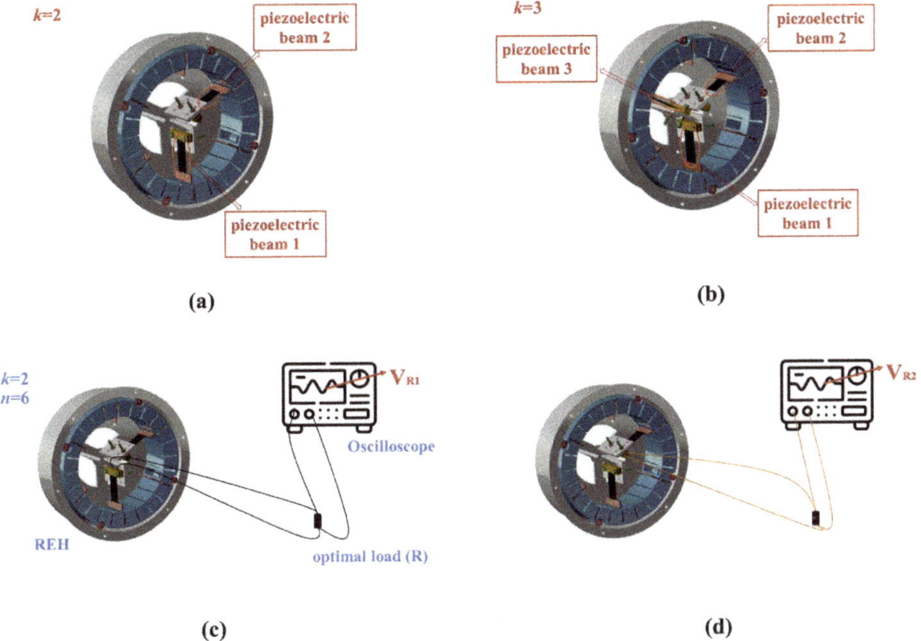

Figure 15. Multi-piezoelectric-beam REH model and output testing method: (**a**) the REH with an internal piezoelectric beam count of $k = 2$; (**b**) the REH with an internal piezoelectric beam count of $k = 3$; (**c**) the measurement of V_R for piezoelectric beam 1; (**d**) the measurement of V_R for piezoelectric beam 2.

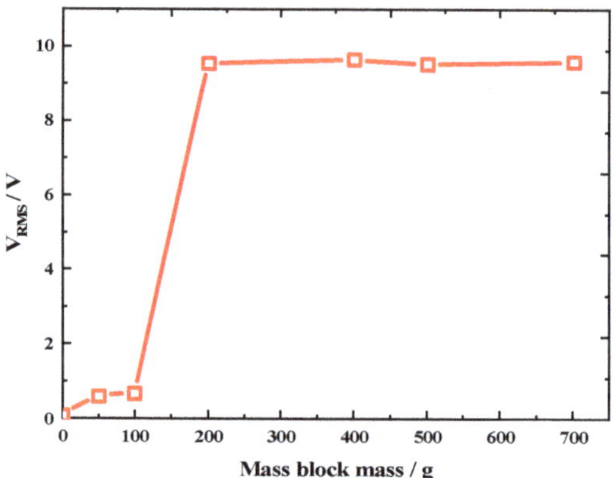

Figure 16. Variation in open-circuit voltage according to the mass block mass (system rotation speed ω = 30 rpm, number of paddles n = 8, number of piezoelectric beams k = 1).

Hence, we investigated the output characteristics of the REH with different numbers of piezoelectric beams when the mass of the quality block was sufficiently large (G = 500 g), as shown in Figure 17. For k = 1, the output power of the REH increased with an increasing number of paddles, reaching a maximum power output of 0.898 mW for REH-8. For k = 2, the REH achieved an output power of 3.052 mW under the same external conditions. With k = 3, the REH achieved an output power of 5.392 mW, with the highest output of an individual beam reaching 2.268 mW. This was achieved by adjusting the relative distance between the top of the piezoelectric beam and the top of the paddle, thereby increasing the contact area during the collision and enhancing the effectiveness of the impact between the paddle and the piezoelectric beam. Thus, it could be concluded that the proposed method of incorporating an array of piezoelectric beams into the REH was feasible for enhancing its output performance.

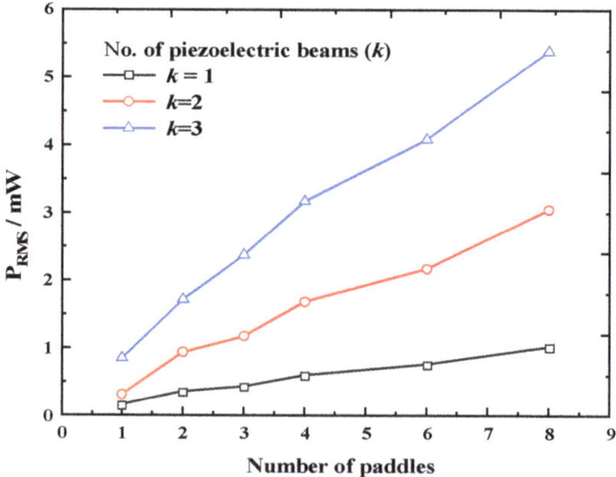

Figure 17. Output power of the REH as a function of the number of paddles under different numbers of piezoelectric beams (system rotation speed ω = 25 rpm).

To evaluate the performance superiority of the designed REH, we compared it with existing energy harvesters, as shown in Table 3. The proposed REH achieved a maximum output power of 5.392 mW, demonstrating a significant advantage in output performance compared to other energy harvesters. Additionally, the maximum power density of the proposed REH reached 4.02 μW/(cm^3 Hz), indicating substantial improvement in output performance compared to the device in our previous work [41], which had a power density of 0.23 μW/(cm^3 Hz).

Table 3. A comparison of the proposed REH and typical REHs at a low rotational speed.

Reference	Number of Piezoelectric Beams	Frequency (Hz)	Power (μW)	Volume [a] (cm^3)	Power Density (μW/(cm^3 Hz))
[29]	1	3	2342	990	0.79
[39]	2	0.79	106	130.8	0.81
[43]	12	3.3	613	45.4	2.28
[41]	1	1.25	141	456.2	0.23
This work	1	0.42	2268	3185.1	1.69
This work	3	0.42	5392	3185.1	4.02

[a] Space demanded by the energy harvester during motion.

5. Conclusions

In this paper, a novel ultra-low-frequency rotational energy harvester based on a double-frequency up-conversion mechanism was proposed, enabling the efficient harvesting of rotational energy in ultra-low-frequency rotational environments (<1 Hz) and demonstrating satisfactory output performance. The proposed rotational energy harvester (REH) consisted of a cylindrical shell with an array of piezoelectric beams and a slider with multiple paddles. During operation, the relative motion between the slider and the shell caused the paddles to periodically strike the piezoelectric beams, inducing self-excited oscillation in the beams (the first frequency up-conversion) and converting mechanical energy into electrical energy through the piezoelectric effect. Additionally, by increasing the number of paddles on the slider and the piezoelectric beams inside the REH, the frequency at which the beams were struck during one rotation cycle could be further increased (the second frequency up-conversion), enhancing the output performance of the REH. Experimental investigations were conducted to study the relationship between the output performance of the proposed REH and parameters such as the number of paddles and the system rotation speed. The data showed that under low-frequency rotation and low-eccentricity conditions, the output performance of the proposed REH could be improved by increasing the number of paddles and the system rotation speed. Furthermore, to further enhance the output performance of the REH, an array of piezoelectric beams was introduced inside the shell, and the output performance of the REH with different numbers of piezoelectric beams was tested. The experimental results revealed that the maximum output power of an individual piezoelectric beam reached 2.268 mW, while the REH with three piezoelectric beams achieved an output power of 5.392 mW, with a power density of 4.02 μW/(cm^3 Hz) under a rotational excitation of 0.42 Hz, exhibiting better output performance compared to similar REHs.

It is worth noting that the main purpose of this study was to demonstrate the advantages of the proposed REH structural design based on a double-frequency up-conversion mechanism in terms of performance enhancement. However, in order to further enhance and evaluate the output performance of the proposed REH, the optimization of the REH design parameters (such as the number of piezoelectric beams, the number of paddles, the eccentricity, and the slider mass) should also be investigated. Furthermore, practical application-related performance testing and the design of energy management circuits must be considered. These issues will be addressed in our future research.

Author Contributions: Conceptualization, N.L.; methodology, N.L. and H.X.; software, N.L.; validation, N.L. and H.X.; formal analysis, N.L. and H.X.; investigation, H.X.; resources, T.L.; data curation, T.L. and C.Y.; writing—original draft preparation, N.L.; writing—review and editing, L.Q. and C.Y.; visualization, T.L.; supervision, L.Q.; project administration, L.Q.; funding acquisition, L.Q. All authors have read and agreed to the published version of the manuscript.

Funding: This work was funded in part by the Knowledge Innovation Program of Shenzhen City (Fundamental Research, Free Exploration) under grant JCY20190809162001746; in part by the Natural Science Foundation of Fujian Province of China (grant no. 2022J01057); and in part by the Guangdong Basic and Applied Basic Research Foundation (2021A1515110504).

Institutional Review Board Statement: Not applicable.

Informed Consent Statement: Not applicable. This study did not involve humans.

Data Availability Statement: Not applicable.

Conflicts of Interest: The authors declare no conflict of interest.

References

1. Sun, F.; Dong, R.; Zhou, R.; Xu, F.; Mei, X. Theoretical and Experimental Investigation of a Rotational Magnetic Couple Piezoelectric Energy Harvester. *Micromachines* **2022**, *13*, 936. [CrossRef] [PubMed]
2. Jiang, J.; Liu, S.; Feng, L.; Zhao, D. A Review of Piezoelectric Vibration Energy Harvesting with Magnetic Coupling Based on Different Structural Characteristics. *Micromachines* **2021**, *12*, 436. [CrossRef]
3. Fu, H.; Yeatman, E.M. A Methodology for Low-Speed Broadband Rotational Energy Harvesting Using Piezoelectric Transduction and Frequency up-Conversion. *Energy* **2017**, *125*, 152–161. [CrossRef]
4. Zhu, Y.; Zhang, Z.; Zhang, P.; Tan, Y. A Magnetically Coupled Piezoelectric–Electromagnetic Low-Frequency Multidirection Hybrid Energy Harvester. *Micromachines* **2022**, *13*, 761. [CrossRef]
5. Cao, L.M.; Li, Z.X.; Guo, C.; Li, P.P.; Meng, X.Q.; Wang, T.M. Design and Test of the MEMS Coupled Piezoelectric–Electromagnetic Energy Harvester. *Int. J. Precis. Eng. Manuf.* **2019**, *20*, 673–686. [CrossRef]
6. Li, G.; Cui, J.; Liu, T.; Zheng, Y.; Hao, C.; Hao, X.; Xue, C. Triboelectric-Electromagnetic Hybrid Wind-Energy Harvester with a Low Startup Wind Speed in Urban Self-Powered Sensing. *Micromachines* **2023**, *14*, 298. [CrossRef] [PubMed]
7. Zhao, H.; Ouyang, H. Erratum: Theoretical Investigation and Experiment of a Disc-Shaped Triboelectric Energy Harvester with a Magnetic Bistable Mechanism. *Smart Mater. Struct.* **2021**, *30*, 095026. [CrossRef]
8. Astigarraga, A.; Lopez-Gasso, A.; Golpe, D.; Beriain, A.; Solar, H.; Del Rio, D.; Berenguer, R. A 21 m Operation Range RFID Tag for "Pick to Light" Applications with a Photovoltaic Harvester. *Micromachines* **2020**, *11*, 1013. [CrossRef]
9. Turkevych, I.; Kazaoui, S.; Shirakawa, N.; Fukuda, N. Potential of AgBiI4rudorffites for Indoor Photovoltaic Energy Harvesters in Autonomous Environmental Nanosensors. *Jpn. J. Appl. Phys.* **2021**, *60*, abf2a5. [CrossRef]
10. Shi, S.; Yue, Q.; Zhang, Z.; Yuan, J.; Zhou, J.; Zhang, X.; Lu, S.; Luo, X.; Shi, C.; Yu, H. A Self-Powered Engine Health Monitoring System Based on L-Shaped Wideband Piezoelectric Energy Harvester. *Micromachines* **2018**, *9*, 629. [CrossRef]
11. Wang, J.; Geng, L.; Zhou, S.; Zhang, Z.; Lai, Z.; Yurchenko, D. Design, Modeling and Experiments of Broadband Tristable Galloping Piezoelectric Energy Harvester. *Acta Mech. Sin. Xuebao* **2020**, *36*, 592–605. [CrossRef]
12. Oh, Y.; Kwon, D.S.; Eun, Y.; Kim, W.; Kim, M.O.; Ko, H.J.; Kang, S.G.; Kim, J. Flexible Energy Harvester with Piezoelectric and Thermoelectric Hybrid Mechanisms for Sustainable Harvesting. *Int. J. Precis. Eng. Manuf. Green Technol.* **2019**, *6*, 691–698. [CrossRef]
13. Fu, H.; Mei, X.; Yurchenko, D.; Zhou, S.; Theodossiades, S.; Nakano, K.; Yeatman, E.M. Rotational Energy Harvesting for Self-Powered Sensing. *Joule* **2021**, *5*, 1074–1118. [CrossRef]
14. Jamadar, V.; Pingle, P.; Kanase, S. Possibility of Harvesting Vibration Energy from Power Producing Devices: A Review. In Proceedings of the 2016 International Conference on Automatic Control and Dynamic Optimization Techniques (ICACDOT), Pune, India, 9–10 September 2016; pp. 496–503. [CrossRef]
15. Zhang, L.; Qin, L.; Qin, Z.; Chu, F. Energy Harvesting from Gravity-Induced Deformation of Rotating Shaft for Long-Term Monitoring of Rotating Machinery. *Smart Mater. Struct.* **2022**, *31*, ac9e2d. [CrossRef]
16. Kayaharman, M.; Das, T.; Seviora, G.; Saritas, R.; Abdel-Rahman, E.; Yavuz, M. Long-Term Stability of Ferroelectret Energy Harvesters. *Materials* **2020**, *13*, 42. [CrossRef]
17. Yu, H.; Zhang, X.; Shan, X.; Hu, L.; Zhang, X.; Hou, C.; Xie, T. A Novel Bird-Shape Broadband Piezoelectric Energy Harvester for Low Frequency Vibrations. *Micromachines* **2023**, *14*, 421. [CrossRef]
18. Akkaya Oy, S. A Piezoelectric Energy Harvesting from the Vibration of the Airflow around a Moving Vehicle. *Int. Trans. Electr. Energy Syst.* **2020**, *30*, 1–13. [CrossRef]
19. Huang, H.H.; Chen, K.S. Design, Analysis, and Experimental Studies of a Novel PVDF-Based Piezoelectric Energy Harvester with Beating Mechanisms. *Sens. Actuators A Phys.* **2016**, *238*, 317–328. [CrossRef]
20. Tommasino, D.; Moro, F.; Bernay, B.; Woodyear, T.D.L.; de Corona, E.P.; Doria, A. Vibration Energy Harvesting by Means of Piezoelectric Patches: Application to Aircrafts. *Sensors* **2022**, *22*, 363. [CrossRef]
21. Xiao, H.; Wang, X.; John, S. A Multi-Degree of Freedom Piezoelectric Vibration Energy Harvester with Piezoelectric Elements Inserted between Two Nearby Oscillators. *Mech. Syst. Signal Process.* **2016**, *68–69*, 138–154. [CrossRef]
22. Qian, Y.; Chen, Y. Research on Multi-Valued Response and Bursting Oscillation of Series Multi-Stable Piezoelectric Energy Harvester. *Eur. Phys. J. Plus* **2022**, *137*, 588. [CrossRef]

23. He, L.; Zhou, J.; Han, Y.; Liu, R.; Tian, X.; Liu, L. Study of a Piezoelectric Energy Harvester in the Form of Vortex Oscillation for Fixed Disturbance Fluid Type. *Rev. Sci. Instrum.* **2022**, *93*, 064705. [CrossRef] [PubMed]

24. Zhang, Y.; Cao, J.; Zhu, H.; Lei, Y. Design, Modeling and Experimental Verification of Circular Halbach Electromagnetic Energy Harvesting from Bearing Motion. *Energy Convers. Manag.* **2019**, *180*, 811–821. [CrossRef]

25. Bao, B.; Wang, Q. Bladeless Rotational Piezoelectric Energy Harvester for Hydroelectric Applications of Ultra-Low and Wide-Range Flow Rates. *Energy Convers. Manag.* **2021**, *227*, 113619. [CrossRef]

26. Qichang, Z.; Yang, Y.; Wei, W. Theoretical Study on Widening Bandwidth of Piezoelectric Vibration Energy Harvester with Nonlinear Characteristics. *Micromachines* **2021**, *12*, 1301. [CrossRef] [PubMed]

27. Wu, Q.; Zhang, H.; Lian, J.; Zhao, W.; Zhou, S.; Zhao, X. Experiment Investigation of Bistable Vibration Energy Harvesting with Random Wave Environment. *Appl. Sci.* **2021**, *11*, 868. [CrossRef]

28. Raja, V.; Umapathy, M.; Uma, G.; Usharani, R. Performance Enhanced Piezoelectric Rotational Energy Harvester Using Reversed Exponentially Tapered Multi-Mode Structure for Autonomous Sensor Systems. *J. Sound Vib.* **2023**, *544*, 117429. [CrossRef]

29. Wang, S.; Yang, Z.; Kan, J.; Chen, S.; Chai, C.; Zhang, Z. Design and Characterization of an Amplitude-Limiting Rotational Piezoelectric Energy Harvester Excited by a Radially Dragged Magnetic Force. *Renew. Energy* **2021**, *177*, 1382–1393. [CrossRef]

30. Chen, C.T.; Su, W.J.; Wu, W.J.; Vasic, D.; Costa, F. Magnetic Plucked Meso-Scale Piezoelectric Energy Harvester for Low-Frequency Rotational Motion. *Smart Mater. Struct.* **2021**, *30*, 105014. [CrossRef]

31. Cao, Y.Y.; Yang, J.H.; Yang, D.B. Dynamically Synergistic Transition Mechanism and Modified Nonlinear Magnetic Force Modeling for Multistable Rotation Energy Harvester. *Mech. Syst. Signal Process.* **2023**, *189*, 110085. [CrossRef]

32. He, L.; Wang, Z.; Wu, X.; Zhang, Z.; Zhao, D.; Tian, X. Analysis and Experiment of Magnetic Excitation Cantilever-Type Piezoelectric Energy Harvesters for Rotational Motion. *Smart Mater. Struct.* **2020**, *29*, 055043. [CrossRef]

33. Guan, M.; Liao, W.H. Design and Analysis of a Piezoelectric Energy Harvester for Rotational Motion System. *Energy Convers. Manag.* **2016**, *111*, 239–244. [CrossRef]

34. Mei, X.; Zhou, R.; Yang, B.; Zhou, S.; Nakano, K. Combining Magnet-Induced Nonlinearity and Centrifugal Softening Effect to Realize High-Efficiency Energy Harvesting in Ultralow-Frequency Rotation. *J. Sound Vib.* **2021**, *505*, 116146. [CrossRef]

35. Janphuang, P.; Lockhart, R.A.; Isarakorn, D.; Henein, S.; Briand, D.; De Rooij, N.F. Harvesting Energy from a Rotating Gear Using an AFM-Like MEMS Piezoelectric Frequency up-Converting Energy Harvester. *J. Microelectromechanical Syst.* **2015**, *24*, 742–754. [CrossRef]

36. Fan, K.; Liu, J.; Wei, D.; Zhang, D.; Zhang, Y.; Tao, K. A Cantilever-Plucked and Vibration-Driven Rotational Energy Harvester with High Electric Outputs. *Energy Convers. Manag.* **2021**, *244*, 114504. [CrossRef]

37. Fan, K.; Wang, C.; Chen, C.; Zhang, Y.; Wang, P.; Wang, F. A Pendulum-Plucked Rotor for Efficient Exploitation of Ultralow-Frequency Mechanical Energy. *Renew. Energy* **2021**, *179*, 339–350. [CrossRef]

38. Zou, H.X.; Zhang, W.M.; Li, W.B.; Gao, Q.H.; Wei, K.X.; Peng, Z.K.; Meng, G. Design, Modeling and Experimental Investigation of a Magnetically Coupled Flextensional Rotation Energy Harvester. *Smart Mater. Struct.* **2017**, *26*, 115023. [CrossRef]

39. Chen, J.; Liu, X.; Wang, H.; Wang, S.; Guan, M. Design and Experimental Investigation of a Rotational Piezoelectric Energy Harvester with an Offset Distance from the Rotation Center. *Micromachines* **2022**, *13*, 388. [CrossRef]

40. Fang, S.; Fu, X.; Du, X.; Liao, W.H. A Music-Box-like Extended Rotational Plucking Energy Harvester with Multiple Piezoelectric Cantilevers. *Appl. Phys. Lett.* **2019**, *114*, 233902. [CrossRef]

41. Xia, H.; Yang, F.; Yang, C.; Qin, L.; Zhang, J. An Eccentric Rotational Energy Harvester Using Liquid as an Energy-Capturing Medium. *Energy Convers. Manag.* **2022**, *265*, 115759. [CrossRef]

42. Machado, L.Q.; Yurchenko, D.; Wang, J.; Clementi, G.; Margueron, S.; Bartasyte, A. Multi-Dimensional Constrained Energy Optimization of a Piezoelectric Harvester for E-Gadgets. *iScience* **2021**, *24*, 102749. [CrossRef] [PubMed]

43. Yang, Y.; Shen, Q.; Jin, J.; Wang, Y.; Qian, W.; Yuan, D. Rotational Piezoelectric Wind Energy Harvesting Using Impact-Induced Resonance. *Appl. Phys. Lett.* **2014**, *105*, 053901. [CrossRef]

Article

Improvement of the Airflow Energy Harvester Based on the New Diamagnetic Levitation Structure

Long Zhang [1], Hang Shao [1], Jiaxiang Zhang [1], Deping Liu [1], Kean C. Aw [2] and Yufeng Su [1,*]

[1] School of Mechanical and Power Engineering, Zhengzhou University, Zhengzhou 450001, China; gffrfzl@126.com (L.Z.); 202022202013965@gs.zzu.edu.cn (H.S.); 15038554491@163.com (J.Z.); ldp@zzu.edu.cn (D.L.)

[2] Department of Mechanical and Mechatronics Engineering, University of Auckland, Auckland 1010, New Zealand; k.aw@auckland.ac.nz

* Correspondence: yufengsu@zzu.edu.cn

Abstract: This paper presents an improved solution for the airflow energy harvester based on the push–pull diamagnetic levitation structure. A four-notch rotor is adopted to eliminate the offset of the floating rotor and substantially increase the energy conversion rate. The new rotor is a centrally symmetrical-shaped magnet, which ensures that it is not subjected to cyclically varying unbalanced radial forces, thus avoiding the rotor's offset. Considering the output voltage and power of several types of rotors, the four-notch rotor was found to be optimal. Furthermore, with the four-notch rotor, the overall average increase in axial magnetic spring stiffness is 9.666% and the average increase in maximum monostable levitation space is 1.67%, but the horizontal recovery force is reduced by 3.97%. The experimental results show that at an airflow rate of 3000 sccm, the peak voltage and rotation speed of the four-notch rotor are 2.709 V and 21,367 rpm, respectively, which are 40.80% and 5.99% higher compared to the three-notch rotor. The experimental results were consistent with the analytical simulation. Based on the improvement, the energy conversion factor of the airflow energy harvester increased to 0.127 mV/rpm, the output power increased to 138.47 mW and the energy conversion rate increased to 58.14%, while the trend of the levitation characteristics also matched the simulation results. In summary, the solution proposed in this paper significantly improves the performance of the airflow energy harvester.

Keywords: push–pull diamagnetic structure; energy harvester; energy conversion rate; electromagnetic

Citation: Zhang, L.; Shao, H.; Zhang, J.; Liu, D.; Aw, K.C.; Su, Y. Improvement of the Airflow Energy Harvester Based on the New Diamagnetic Levitation Structure. *Micromachines* **2023**, *14*, 1374. https://doi.org/10.3390/mi14071374

Academic Editor: Erwin Peiner

Received: 16 June 2023
Revised: 29 June 2023
Accepted: 1 July 2023
Published: 4 July 2023

1. Introduction

With the rapid development of wireless sensor networks [1], microelectromechanical systems [2], and the Internet of Things [3], more and more demands are being placed on portable power supplies. Traditional chemical batteries cannot provide a long-term and stable power supply [4] due to their low energy density, inability to be recycled, and environmental hazards [5]. The collection of energy from the environment and its conversion into electrical energy for use in electronic devices has been the subject of much research to meet the power requirements of these devices [6]. Energy harvesting technologies involve conversion mechanisms such as electromagnetic [7,8], triboelectric [9,10], electrostatic [11,12], magnetostrictive [13,14], thermoelectric [15,16], piezoelectric [17,18] and photovoltaic [19,20]. Energy harvesters typically use one or more conversion mechanisms to convert energy from nature: tides, vibrations, air currents, heat, etc., into electrical energy.

Airflow is a kind of widespread source of clean energy in nature, and it has many sources and the widest range of applications. Xin et al. [21] propose a two-dimensional airflow energy harvester that collects vibrational energy from the airflow in all directions via cantilever beams and piezoelectric tubes, achieving an output power of 0.353–0.495 mW at wind speeds of 8 m/s. Wang et al. [22] proposed a non-contact piezoelectric wind energy harvesting device to harvest wind-excited vibration energy with an adjustable structure to suit different wind speeds, with a maximum output of 1.438 mW at a wind speed of 40 m/s. Wang et al. [23] proposed an improved method based on a flapping airflow energy harvester, introducing flexible wing sections to increase the output power, which reached about 930 mW at a wind speed of 9 m/s when the flexible section was 5 cm. In all these articles, the output performance of the airflow energy harvesters is not high, while the gap between the collected energy and the electrical energy output is too large to achieve a high energy conversion rate.

Diamagnetic levitation was first investigated experimentally by Cansiz and Hull [24] in 2004. The non-contact nature of the diamagnetic levitation structure avoids friction between moving parts, so it can be used in airflow energy harvesters to achieve high output performance. In our previous work [25], a push–pull diamagnetic levitation structure was reported, and the new diamagnetic levitation structure was used to harvest airflow energy. The airflow energy harvester [26] operated with good stability and environmental adaptability.

This paper focuses on improving the airflow energy harvester based on the new diamagnetic levitation structure to enhance output performance and energy conversion rate. A centrosymmetric rotor is proposed to address the shortcomings of the three-notch rotor. Simulation models are built in COMSOL to compare and verify various floating rotors' output performance and determine the optimum rotor parameters. A joint COMSOL and MATLAB simulation was carried out to compare the levitation characteristics of the new rotor with those of the three-notch rotor. Experimental prototypes and platforms have been built based on theoretical and simulated models, and the results show that the airflow energy harvester with the new rotor has higher output performance and better levitation characteristics.

2. Theoretical Analysis

2.1. Analysis of Fundamentals

The three-dimensional model of the airflow energy harvester is shown in Figure 1. It includes the pushing magnet, the upper highly oriented pyrolytic graphite (HOPG) sheet, top coils, the floating rotor, bottom coils, the lower highly oriented pyrolytic graphite sheet, the pulling magnet, and two airflow nozzles. All the structures except the nozzles are arranged coaxially. The two nozzles are placed symmetrically around the vertical axis of the energy harvester and located in the central horizontal plane of the floating rotor.

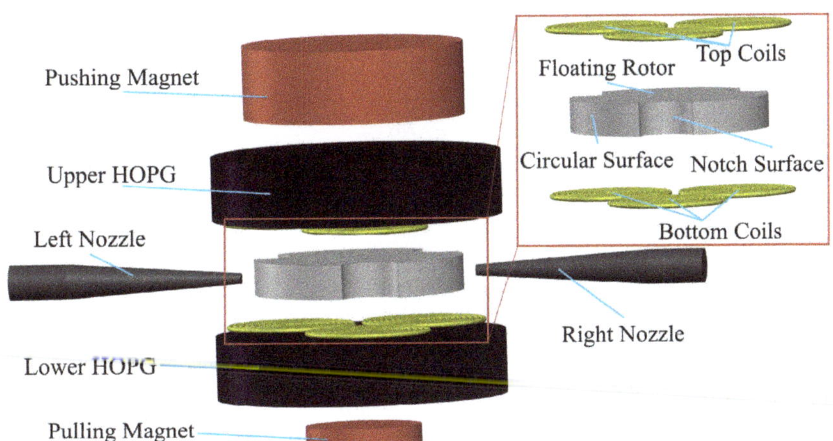

Figure 1. 3D schematic of the airflow energy harvester.

The push–pull diamagnetic levitation structure is shown in Figure 2a, the floating rotor has the same magnetization direction as the pushing magnet and pulling magnet, so the floating rotor is subject to their magnetic attraction F_{Pus} and F_{Pul}. Diamagnetic force F_{Up} and F_{Low} are exerted on the floating rotor from the upper and lower HOPG sheets. In Figure 2b, the axial resultant force of the floating rotor can be expressed as,

$$F_R = F_{Pus} + F_{Low} - F_{Pul} - G - F_{Up} \tag{1}$$

where G represents the gravity of the floating rotor, and F_R denotes the axial resultant force. The potential energy of the floating rotor can be expressed by Equation (2).

$$U = -\vec{M} \cdot \vec{B} + mgz = -\vec{M}\left(\vec{B}_{Pus} - \vec{B}_{Pul}\right) + mgz = -M(B_{Pus} - B_{Pul}) + mgz \tag{2}$$

where M is the magnetic dipole moment of the floating rotor, B_{Pus} denotes the magnetic flux density of the pushing magnet, B_{Pul} shows the magnetic flux density of the pulling magnet, m shows the mass of the floating rotor, and z shows the distance from the ground. Equation (3) can be obtained by substituting the diamagnetic influence exponents [27] C_z and C_r into Equation (2).

$$U = -M\left\{ \begin{array}{l} (B_{Pus0} - B_{Pul0}) + \left[(B_{Pus} - B_{pul})' - \frac{mg}{M}\right]z + \frac{1}{2}(B_{Pus} - B_{Pul})''z^2 \\ + \frac{1}{4}\left[\frac{(B_L - B_p)'^2}{2(B_{L0} - B_{P0})} - (B_{Pus} - B_{pul})''\right]r^2 + \cdots \end{array} \right\} + C_z z^2 + C_r r^2 \tag{3}$$

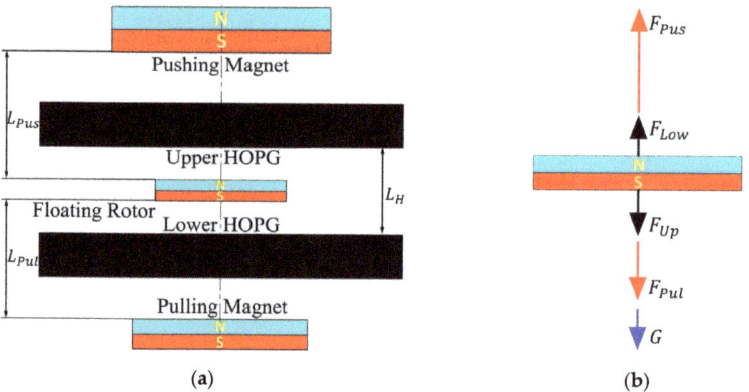

Figure 2. Schematic diagram of the new diamagnetic levitation structure: (**a**) structure schematic diagram; (**b**) force diagram of the floating rotor.

The formula for the stability of the floating rotor that can be derived from Equations (3)–(5) are the equations for vertical and horizontal stability.

$$K_v \equiv C_z - \frac{1}{2}M(B_{Pus} - B_{Pul})'' > 0 \text{ (Vertical stability)} \tag{4}$$

$$K_h \equiv C_r + \frac{1}{4}M\left\{(B_{Pus} - B_{Pul})'' - \frac{m^2 g^2}{2M^2(B_{L0} - B_{P0})}\right\} > 0 \text{ (Vertical stability)} \tag{5}$$

Compared with the old structure [28], the push–pull diamagnetic levitation structure only introduces a pulling magnet, but has the advantages of multiple levitation equilibrium points, multiple maximum monostable levitation spaces, a large increase in horizontal recovery force, and more stable axial force distribution. The airflow energy harvester with the push–pull diamagnetic levitation structure has better output performance and a wider application range.

2.2. Analysis of the Floating Rotor

The schematic diagram of the three-notch rotor is shown in Figure 3, with its radius, thickness, notch radius and central hole radius being 9 mm, 3 mm, 2.5 mm, and 1 mm, respectively. The floating rotor rotates around the central axis under the driving of two symmetrical and nozzle airflow. When the airflow collides with the floating rotor, the motion state of the airflow will be changed. According to the momentum theorem, the force between the floating rotor and the airflow is generated, which changes the velocity direction and magnitude of the airflow, and the floating rotor starts to rotate under the action of this force.

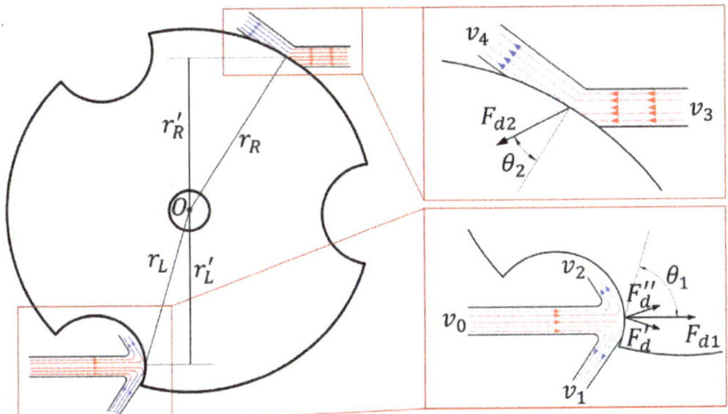

Figure 3. Diagram of the driving forces on the floating rotor.

When the airflow collides with the floating rotor, the motion state of the airflow can be simplified as shown in Figure 3, where red represents the airflow before the collision with the rotor and blue represents the airflow after the collision with the rotor. The driving forces at the notch surface and the circular surface are F_{d1} and F_{d2}, respectively.

$$F_{d1} = F'_d + F''_d = \iiint_{V_1} \rho v_1 dV - \iiint_{V_1} \rho v_0 dV + \iiint_{V_2} \rho v_2 dV - \iiint_{V_2} \rho v_0 dV \tag{6}$$

$$F_{d2} = \iiint_V \rho v_4 dV - \iiint_V \rho v_3 dV \tag{7}$$

Connecting the collision point of the airflow and the center of the floating rotor, θ represents the angle between the driving force and the line connecting the two. When θ is closer to $90°$, the driving effect on the rotor is more prominent, while the radial force on the rotor is smaller. It can be seen from Figure 3 that the angle θ_1 is greater than θ_2, and the tangential component of the driving force F_{d1} is larger, while the radial component of the driving force F_{d2} is larger.

According to the relative positions of the nozzle and itself, the floating rotor can be divided into three parts, red, green, and blue areas shown in Figure 4a, each of which is $120°$. In a whole rotation of the floating rotor, the airflow from the right-side nozzle will blow to three areas in turn. Therefore, the motion state of the floating rotor will undergo three periodic changes in each rotation. As shown in Figure 4b, each period can be divided into four stages according to the different positions of the left and right-side nozzles. During the four stages of each period, the relative position between the floating rotor and the nozzles changes sequentially from position 1 to position 4, while the airflow is directed towards the floating rotor's yellow, red, green, and cyan areas, respectively. Furthermore, the angle α of the yellow and green areas is $32°$, and the angle β of the red and cyan areas is $28°$.

$$F_H = F_r = F_{Ldr} + F_{Rdr} \tag{8}$$

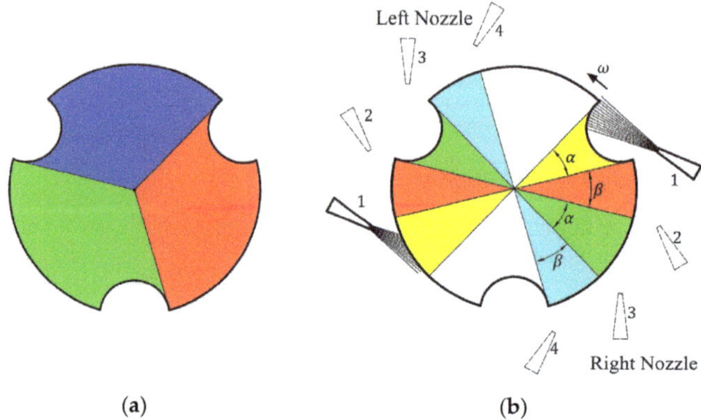

(a)   (b)

Figure 4. The floating rotor area division: (**a**) rotation period division; (**b**) motion state period division.

The first stage is shown in Figure 5. In Figure 5a, the driving force of the airflow from the right side and left nozzles on the floating rotor are F_{Rd} and F_{Ld}, respectively. The two forces are decomposed into two tangential forces F_{Rdt} and F_{Ldt}, and radial forces F_{Rdr} and F_{Ldr}. As can be seen from Figure 3, the two radial forces cannot completely cancel each other out; hence, the resultant force of the two is F_r, which makes the floating rotor unbalanced in the horizontal plane and thus begins to offset. When the pushing magnet, the pulling magnet, and the floating rotor are not coaxial, the pushing magnet and pulling magnet exert a horizontal magnetic force on the floating rotor, and the sum of the two horizontal magnetic forces is the horizontal recovery force F_H of the floating rotor. In Figure 5b, the floating rotor begins to offset under the action of radial force, and with the increase in the offset, the horizontal recovery force F_H also increases. As expressed in Equation (8), when it reaches O', the value of the horizontal recovery force F_H is equal to that of the radial force F_r, the floating rotor reaches its equilibrium in the horizontal direction, and the offset also reaches the maximum value. O' is denoted as the maximum right side offset point of the floating rotor, where the floating rotor continues to rotate and enters the next stage, as shown in Figure 6.

$$F_{Ldr} = F_{Rdr} \tag{9}$$

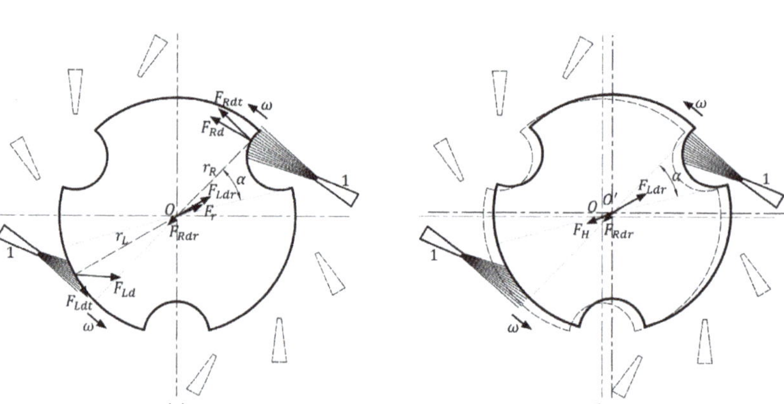

(a)   (b)

Figure 5. First stage: (**a**) the floating rotor at point O; (**b**) the floating rotor at point O'.

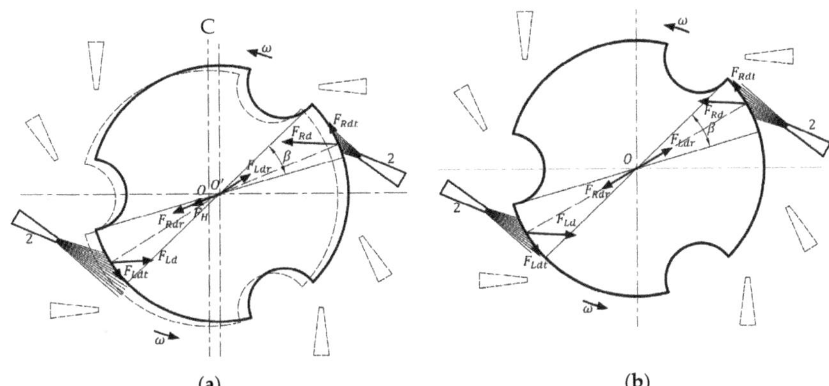

Figure 6. Second stage: (**a**) the floating rotor at point O'; (**b**) the floating rotor at point O.

In Figure 6a, at the beginning of the second stage, since the floating rotor is still located at the maximum right-side offset point, the right-side nozzle is closer to the floating rotor, so the right-side driving force F_{Rd} is significantly greater than the left driving force F_{Ld}. In the horizontal direction, the floating rotor will return to the original central O point under the combined action of the horizontal recovery force F_H, the right-side radial force F_{Rdr} and the left radial force F_{Ldr}. In Figure 6b, when the floating rotor returns to point O, the left and right-side nozzles are at an equal distance from the floating rotor, the radial and tangential components of the left and right-side driving forces F_{Ld} and F_{Rd} are equal, as shown in Equation (9). The floating rotor enters the third stage after $30°$ rotation around O.

In the third stage, the airflow from the left nozzle blows toward the notch surface, while the right nozzle blows toward the circular surface, which corresponds to the rotation of the floating rotor by $180°$ relative to the first stage. At this time, the direction of the horizontal resultant force F_r is exactly opposite to the horizontal resultant force in the first stage, the floating rotor will be offset to the left, finally reaching the maximum left offset point O''. The floating rotor will enter the fourth stage with another $30°$ rotation. The fourth stage is like the second one in that the floating rotor returns to the initial point O again under the action of the horizontal recovery force F_H and the two radial forces. The entire period is the completion of the rotation of the floating rotor before entering the next rotation period.

During one rotation cycle, the floating rotor is offset to the left and right in the horizontal plane in turn because radial forces cannot be fully counteracted. After reaching a maximum speed of 20,000 rpm, the floating rotor undergoes more than 900 periodic offsets per second. Such a high frequency of periodic offset consumes substantial energy, reducing the energy conversion rate. If the floating rotor deflection is to be eliminated, it must be ensured that the airflow from both nozzles is blowing simultaneously toward the circular surface or the notch surface. The three-notch rotor cannot eliminate the offset due to its structure. When the number of notches is even, the rotor has a centrosymmetric structure, as shown in Figure 7. In both the four-notch rotor and the six-notch rotor, when the nozzles are at position 1, the radial component of the airflow flow can be completely canceled out because the left and right nozzles blow simultaneously onto the circular surface. When the nozzle is at position 2, both the left and right nozzles blow towards the notch surface, and the two radial directions can still cancel each other so that the radial forces on the rotor are balanced. With a centrally symmetrical rotor, both nozzles blow simultaneously onto the notch or circular surface regardless of the rotor rotation angle, eliminating rotor drift and thus achieving higher energy conversion rates, which can be considered an improved solution for airflow energy harvesters.

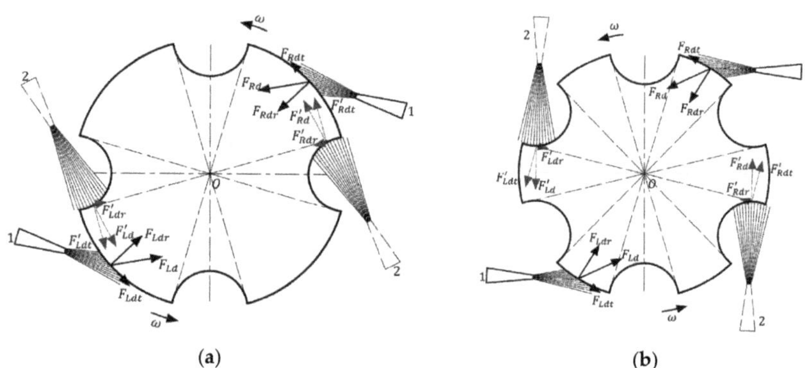

(a)   (b)

Figure 7. The centrosymmetric rotors: (**a**) the four-notch rotor; (**b**) the six-notch rotor.

3. Simulation Analysis of Centrosymmetric Rotor

3.1. Analysis of the Output Performance

In the airflow energy harvester, two factors influence the magnitude of the peak voltage. One is the rotation speed ω of the floating rotor. Higher speed led to higher induction electromotive force in the coil. The other one is the energy conversion factor ρ_{ECF}. The quotient of the peak voltage and the speed of the rotor is the value of the energy conversion factor. The peak voltage E_P can, therefore, be expressed by Equation (10).

$$E_P = \omega \rho_{ECF} \tag{10}$$

The energy conversion part of the energy harvester consists of two shaped coils made up of three circular coils connected in series, which are arranged above and below the floating rotor, as shown in Figure 8a. The simulation is modeled in COMSOL Multiphysics 5.6 as shown in Figure 8b, with the structure parameters selected from Table 1. The pushing magnet, pulling magnet, and floating magnet are made of NdFeB 52 and have a maximum magnetic energy of 400 kA/m^3. The simulation model shown in Figure 8 is used to simulate the output voltages of two-notch, three-notch, four-notch, and six-notch rotors. The speed of three centrosymmetric rotors and the three-notch rotor is set to 20,000 rpm and the simulation time is set to the time required for half a revolution. Figure 9 shows the induction electromotive force corresponding to the four rotors, respectively.

Table 1. Parameters of the push–pull airflow energy harvester.

Parameter	Value I Material
Pushing magnet, Pulling magnet, Floating rotor	NdFeB-52
Pushing magnet	∅19 × 6.35 (mm)
Pulling magnet	∅10 × 2 (mm)
Radius of the floating rotor	9 (mm)
Thickness of the floating rotor	3 (mm)
Radius of the central bore of the floating rotor	1 (mm)
Radius of the notches of the floating rotor	2.5 (mm)
Outer diameter of circular coils	11.5 (mm)
Inner diameter of circular coils	0 (mm)
Wire diameter for circular coils	0.1 (mm)
L_d	0.5 (mm)

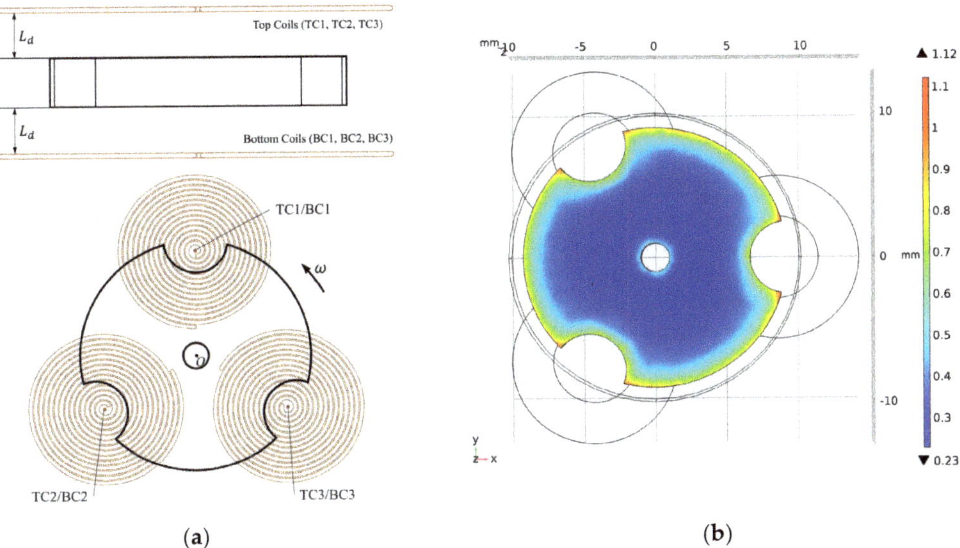

(**a**)   (**b**)

Figure 8. The energy conversion part: (**a**) schematic diagram of position relation of the three-notch rotor and coils; (**b**) voltage simulation model by COMSOL.

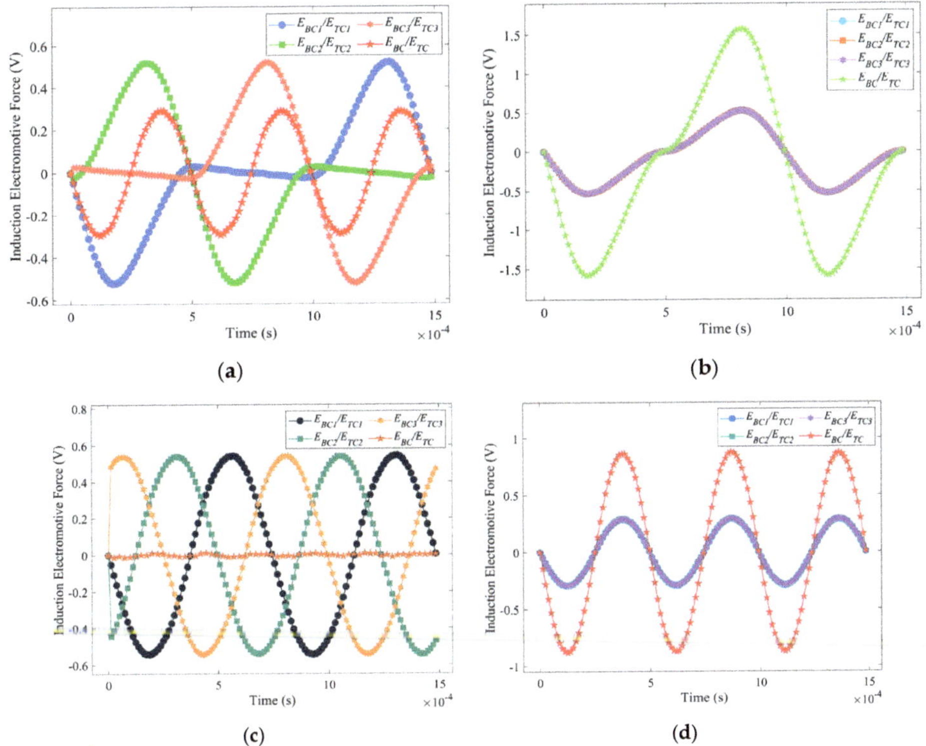

(**a**)   (**b**)

(**c**)   (**d**)

Figure 9. Electromotive force and total voltage of the three coils of the four rotors: (**a**) two-notch rotor; (**b**) three-notch rotor; (**c**) four-notch rotor; (**d**) six-notch rotor.

As seen in Figure 9a,c, the phase of the induced electric potential generated by the three coils to be different, which causes the voltages in the three coils to cancel each other out, and the total voltage E_{TC} (E_{BC}) in the top coils (bottom coils) is less than the sum of the voltages of the three coils, as expressed in Equation (11). Moreover, the value of E_{TC} (E_{BC}) is always zero when using the four-notch rotor. In contrast, as seen in Figure 9b,d, the E_{TC} (E_{BC}) of both the three-notch and six-notch rotors are three times the electromotive force in a single coil, as expressed in Equation (12). Furthermore, the total voltage of the three-notch rotor is significantly higher than that of the six-notch rotor by a factor of approximately two. The energy conversion factors ρ_{ECFS2}, ρ_{ECFS3}, ρ_{ECFS4} and ρ_{ECFS6} of the four rotors in a single coil are calculated from Equation (10), and the values of them are 0.0260 mV/rpm, 0.0263 mV/rpm, 0.0272 mV/rpm, and 0.0146 mV/rpm, respectively. According to Equation (13), the centrosymmetric rotors may have a higher output performance. The coil arrangement of the three-notch rotor is not suitable for the centrosymmetric rotors, which results in their low total voltage.

$$E_{BC}(E_{TC}) < E_{BC1}(E_{TC1}) + E_{BC2}(E_{TC2}) + E_{BC3}(E_{TC3}) \tag{11}$$

$$E_{BC}(E_{TC}) = E_{BC1}(E_{TC1}) + E_{BC2}(E_{TC2}) + E_{BC3}(E_{TC3}) \tag{12}$$

$$\rho_{ECFS4} > \rho_{ECFS3} > \rho_{ECFS2} > \rho_{ECFS6} \tag{13}$$

With the centrosymmetric rotors, the existing coil arrangement must be changed to avoid the induction electromotive force canceling each other out. The improved coil arrangement is shown in Figure 10; several circular coils are fixed on each of the upper and lower HOPG plates that are connected in series and tangent to each other, with the number of them equal to the number of notches of the floating magnet. Using these improved coil arrangements, the induction voltages of the four rotors were simulated in COMSOL Multiphysics 5.6 and plotted in Figure 11. In Figure 11a, the peak electromotive force in the single coil E_{SC} of the two-notch rotor, three-notch rotor, four-notch rotor, and six-notch rotor are 0.3392 V, 0.4344 V, 0.5415 V, and 0.2839 V, respectively. The output voltage E is calculated according to Equation (14), where N is the number of notches in each rotor. In Figure 11b, the peak output voltages of four rotors are 0.6784 V, 2.6064 V, 4.332 V, and 3.4068 V, respectively. Both the total output voltage and the electromotive force in the individual coils are at their maximum when using the four-notch rotor. As a result, the four-notch rotor has the highest energy conversion factor among the four rotors, boosting the output voltage by 1.7256 V compared to the three-notch rotor.

$$E = E_{TC} + E_{BC} = 2NE_{SC} \tag{14}$$

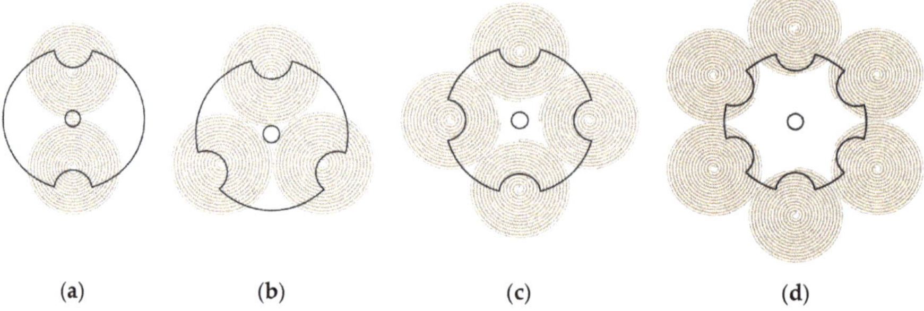

(a)   (b)   (c)   (d)

Figure 10. The improved coil arrangement of the centrosymmetric rotors: (**a**) two-notch rotor; (**b**) three-notch rotor; (**c**) four-notch rotor; (**d**) six-notch rotor.

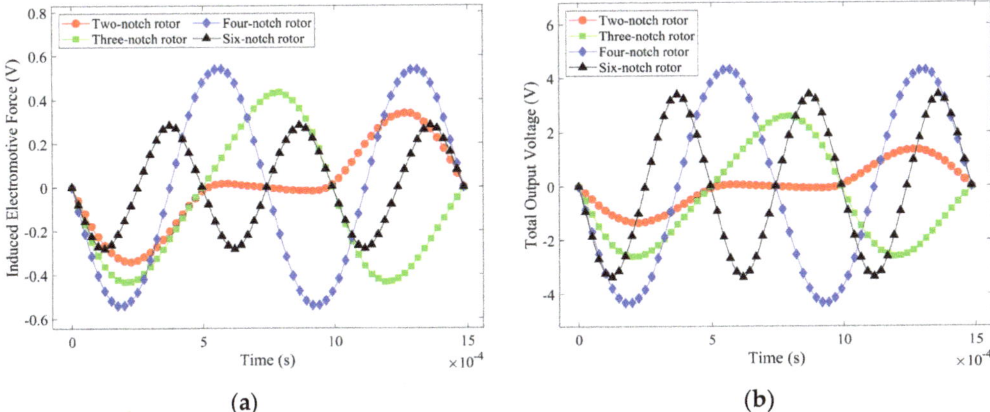

Figure 11. Electromotive force in a single coil and total output voltage of the four rotors with improved coil arrangements: (**a**) induction electromotive force in a single coil; (**b**) total output voltage.

The simulation gives a resistance of 2.42 Ω for a single circular coil. The power of four rotors was calculated from the total output voltage and resistance, as shown in Figure 12. The four-notch rotor has the highest output power of 485 mW. In summary, the four-notch rotor is the best in terms of both energy conversion factor and output power.

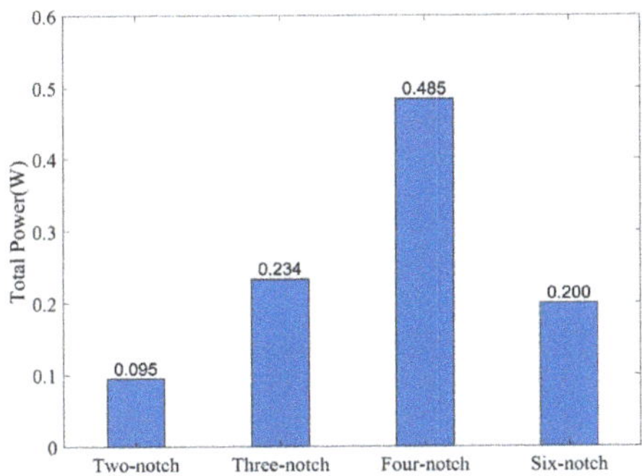

Figure 12. The output power of the four rotors.

3.2. Analysis of Levitation Characteristics

The levitation characteristics of the three-notch and four-notch rotors were calculated using a joint MATLAB and COMSOL simulation. The magnetic and diamagnetic forces of the four-notch rotor are smaller compared to the three-notch rotor due to the lower magnetic induction strength, which makes the axial combined forces and the potential energy of the two types of floating rotor follow approximately the same trend, with slight differences, as shown in Figure 13a,b. This means that there are also differences in the levitation characteristics [25] of the two rotors.

The axial magnetic spring stiffness of the floating rotor represents the ability of the energy harvester to cope with external axial excitation. The larger the axial spring stiffness, the higher the axial stability. As seen in Figure 14, the axial magnetic spring stiffness of the energy harvester is significantly increased when using the four-notch rotor. At L_{Pus} values of 59 mm, 58.5 mm, 58 mm, 57.5 mm, and 57 mm, the axial magnetic spring stiffness increases by an average of 7.886%, 8.328%, 9.915%, 10.939%, and 11.263%, respectively. The overall average increase in axial magnetic spring stiffness is 9.666%.

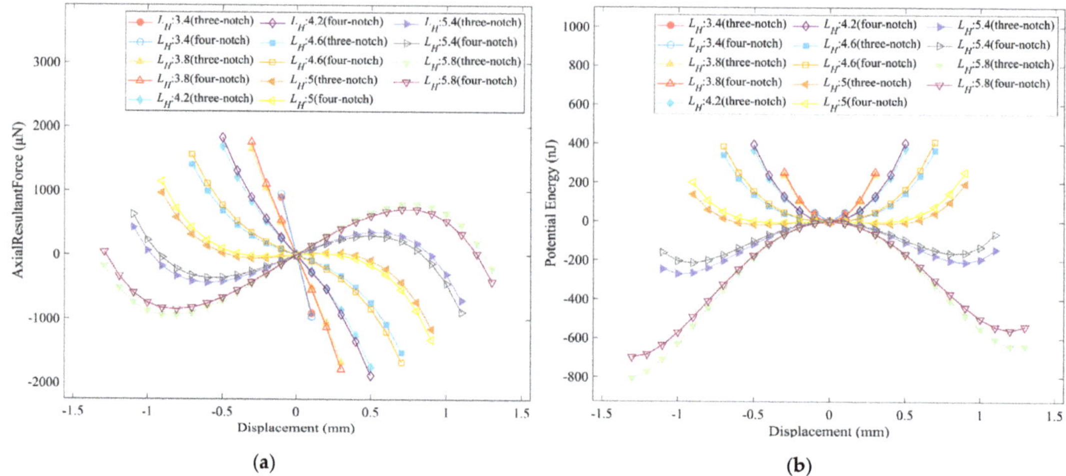

(a)

(b)

Figure 13. The axial resultant force and potential energy of the three-notch rotor and the four-notch rotor: (**a**) the axial resultant force; (**b**) the potential energy.

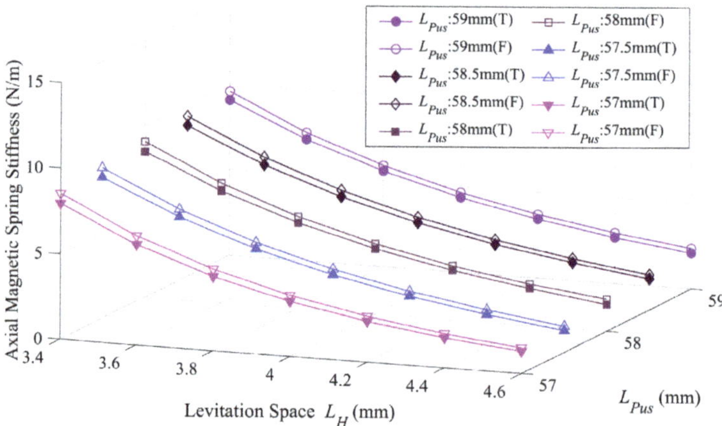

Figure 14. Comparison of the axial magnetic spring stiffness of the three-notch rotor and the four-notch rotor.

In Figure 15, each value of L_{Pus} represents a levitation point, and at each levitation point, there is a maximum monostable levitation space L_{Hmax}. When using the four-notch rotor, L_{Pus} takes on a larger range of values and more levitation equilibrium points can be selected. For the same value of L_{Pus}, the maximum monostable levitation space of the four-notch rotor is somewhat enhanced, with an overall average enhancement of 1.67%. Furthermore, the maximum monostable levitation space's maximum value is increased from 4.88 mm to 5.14 mm. The larger maximum monostable levitation space facilitates observation of the levitation state of the floating rotor and reduces the possibility of friction between the floating rotor and the coil due to axial excitation.

As shown in Figure 16a, the horizontal magnetic force on the four-notch rotor has decreased to some extent, but the reduction is not significant, with the horizontal component of the magnetic force of the pushing magnet decreasing by an average of 4.56% and the horizontal component of the magnetic force of the pulling magnet decreasing by an average of 2.71%. In Figure 16b, the horizontal recovery force for the four-notch rotor has the same trend, with an average reduction of 3.97% compared to the three-notch rotor. The reduction in horizontal recovery force does not affect the performance of the energy harvester as the four-notch rotor does not offset during operation. Moreover, the reduction is within 5%, does not significantly impact horizontal stability, and can cope with horizontal excitation in the operating environment.

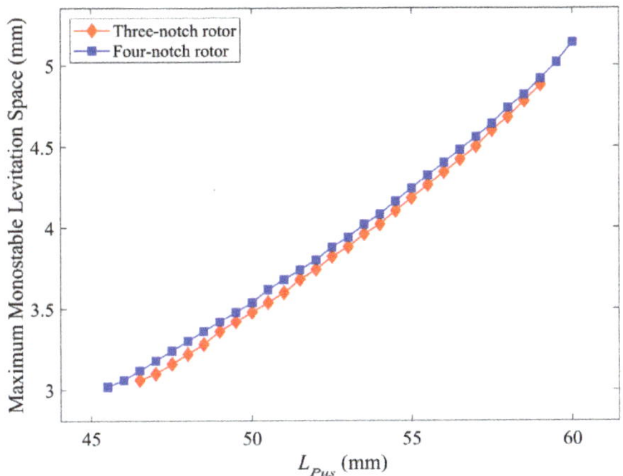

Figure 15. Comparison of maximum monostable levitation space at multiple levitation points for the three-notch rotor and the four-notch rotor.

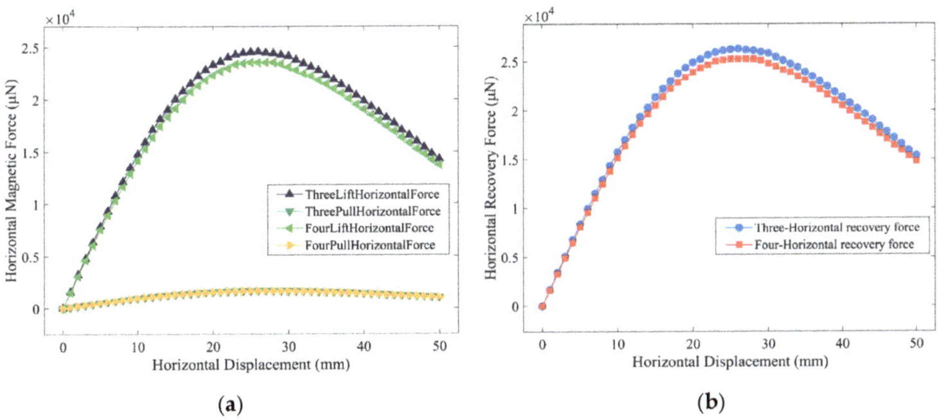

Figure 16. The horizontal component of magnetic force and horizontal recovery force for the three-notch rotor and the four-notch rotor: (**a**) comparison of the horizontal component of magnetic forces; (**b**) comparison of horizontal recovery forces.

4. Experimental Verification

4.1. Levitation Characteristics Test Experiments

The experimental platform of the improved airflow energy harvester is built according to Figure 1, as shown in Figure 17, with the same structural parameters as Table 1. The experiments used two 1.5-mm thick floating rotors stacked together in place of a 3-mm thick floating rotor. The prototype is fixed to a non-magnetic acrylic plate with the aid of multiple adjustment tables and connections printed from photosensitive resin to ensure that the pushing magnet, the pulling magnet, the upper HOPG, and the lower HOPG can be arranged coaxially. Moreover, that the adjustment tables adjust the axial distance between them. In this experiment, only the right-side nozzle was used, and the airflow was regulated by a single flow controller (MFC300). A laser displacement sensor (LK-G80) is placed on the left-hand adjustment table to measure the horizontal displacement of the floating rotor. LK-G80 is connected to the controller (LK-G3001) to display the displacement data, which are recorded by a PC connected to the controller via USB.

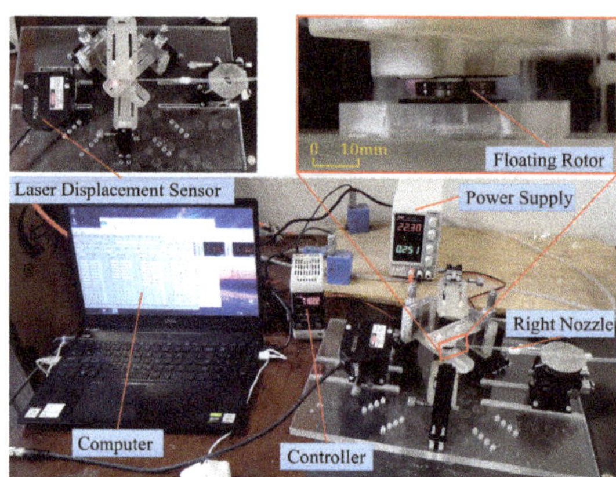

Figure 17. Experimental platform for the levitation characteristics of the airflow energy harvester.

Experiments were first carried out on the maximum monostable levitation space at multiple levitation equilibrium points for the three-notch rotor and the four-notch rotor. The axial positions of the pushing magnet, the pushing magnet, and the two HOPG sheets are changed so that the floating rotor is at different levitation equilibrium points. The thickness of each component and the axial distance between them and the base plate are measured using Vernier calipers. The distance between the upper and lower HOPG plates was calculated from the measured data to be the maximum monostable levitation space and recorded in Table 2. As can be seen from Table 2, as L_{Pus} decreases, the rotor can be kept levitation by reducing L_{Pul}. When the value of L_{Pus} is reduced from 55.365 mm to 51.365 mm, the maximum monostable levitation space L_{Hmax3} is reduced from 4.73 mm to 3.89 mm for the three-notch rotor and L_{Hmax4} is reduced from 4.81 mm to 3.97 mm for the four-notch rotor. Meanwhile, L_{Hmax4} is consistently greater than L_{Hmax3}, with an average improvement of 2.05%, in line with the simulation results in Section 3.

Table 2. Experimental data on maximum monostable levitation space.

Number	L_{Pus} (mm)	L_{Pul3} (mm)	L_{Pul4} (mm)	L_{Hmax3} (mm)	L_{Hmax4} (mm)
1	55.365	46.885	47.385	4.73	4.81
2	54.365	41.885	42.385	4.53	4.63
3	53.365	38.385	39.285	4.31	4.41
4	52.365	35.385	35.885	4.09	4.17
5	51.365	33.485	33.885	3.89	3.97

Both rotors are tested for horizontal recovery force when L_{Pus} is 55.365 mm. The airflow from the nozzle is increased from 0 sccm to 3000 sccm in 200-sccm increments. The horizontal displacement of the floating rotor is measured by a laser displacement sensor (LK-G80, Keyence); the larger the horizontal displacement value, the smaller the horizontal recovery force. The data are further processed and plotted in Figure 18.

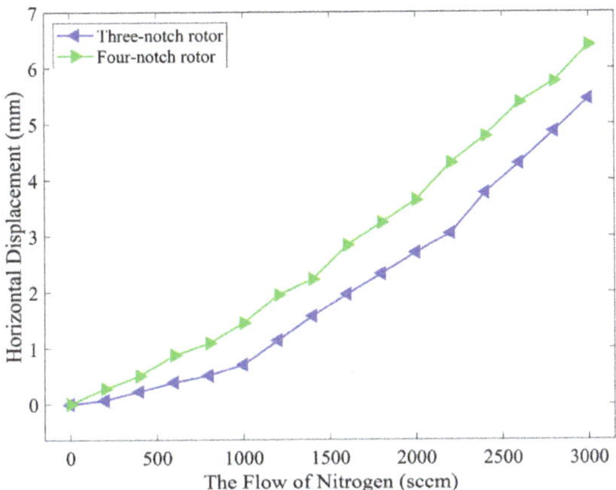

Figure 18. Horizontal displacement of the three-notch rotor and the four-notch rotor at different airflow rates.

In Figure 18, the horizontal displacement of the four-notch rotor is greater than that of the three-notch rotor at any airflow rate. The horizontal displacement reaches a maximum at an airflow rate of 3000 sccm, when the maximum horizontal displacement is 5.45 mm and 6.45 mm for the three-notch rotor and the four-notch rotor, respectively, with an increase of 17.61%. The experimental results agree with the analysis in that the horizontal recovery force of the four-notch rotor is reduced.

4.2. Output Performance Test Experiments

The experimental platform for testing the output performance is shown in Figure 19. Connected to a computer, two flow controllers (MFC300) are used to regulate the airflow of the left and right nozzles. Tektronix oscilloscope is used to display the output waveform of the energy harvester. The coil arrangements of the three-notch and four-notch rotors are fixed on the upper and lower HOPG sheets in turn. The airflow rate of the two nozzles was set to 3000 sccm, and the output performance of the energy harvester prototypes with the three-notch rotor and four-notch rotor was tested. The induced voltage in the coil is generated by the varying magnetic induction strength. Each time the notch surface of the rotor passes over the coil, the voltage in the coil peaks and troughs once. This means that for one revolution of the rotor, the number of peaks and troughs in the output voltage is equal to the number of notches in the rotor. Thus, the rotational speed of the rotor can be calculated from the voltage waveform. The output voltage waves are shown in Figure 20.

In Figure 20a, the red dots represent the voltage of the three-notch rotor, the blue dots represent the voltage of the four-notch rotor. The peak voltage is 1.924 V for the three-notch rotor and 2.709 V for the four-notch rotor, with approximately 40.80% increase. The voltage waveform displayed on the oscilloscope gives a maximum speed of 20,160 rpm for the three-notch rotor and 21,367 rpm for the four-notch rotor. The rotation speed increase is about 5.99% because air resistance is increasing with rotor speed at the same time. The oscilloscope voltage wave of the four-notch rotor is shown in Figure 20b. The total resistance of the six coils connected in series was 18.9 Ω and the total resistance of the eight coils connected in series was 26.5 Ω. For the energy harvesters with three-notch rotor and four-notch rotor, the output power increased from 97.93 mW to 138.47 mW with an increase of 41.40%. Furthermore, the energy conversion factor increased from 0.095 mV/rpm to 0.127 mV/rpm with an increase of 32.68%. The sum of the kinetic energy of the airflow from the left and right nozzles per second is the input power of the experimental prototype, which is approximately 238.16 mW. The energy conversion rate of the energy harvester is equal to the ratio of the output power to the input power, which is 41.12% for the three-notch rotor and 58.14% for the four-notch rotor, with an increase of 41.39%. The four-notch rotor can output higher voltages and power while keeping the input power constant, thus significantly increasing the energy conversion rate. According to Equation (10), the output voltage of the airflow energy harvester is related to the rotation speed and the energy conversion factor. The four-notch rotor has a higher rotational speed and a higher energy conversion factor, so it can output higher power and the energy conversion rate is significantly increased.

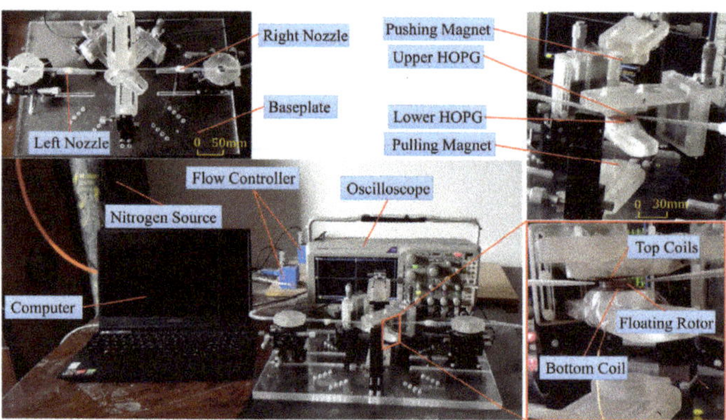

Figure 19. Experimental platform for the airflow energy harvester.

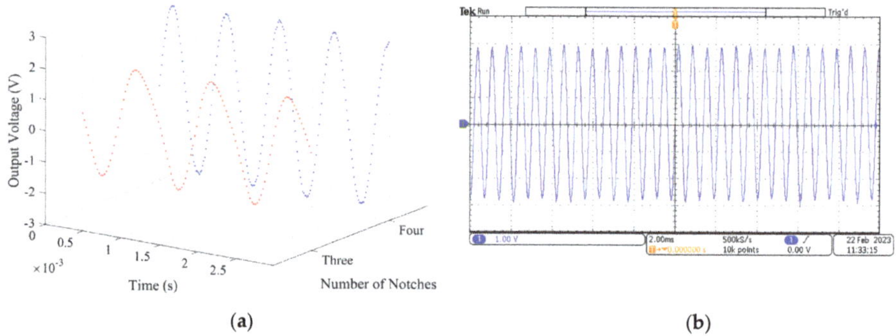

(**a**)   (**b**)

Figure 20. The output voltage of the three-notch rotor and the four-notch rotor: (**a**) voltage data of the two rotors; (**b**) voltage wave of the four-notch rotor.

The four-notch rotor significantly increases the power output and energy conversion rate of the energy harvester, which means that more electronic devices can be loaded and the collected wind energy can be converted into electricity more efficiently. The energy harvester can harvest the wind energy in the environment more effectively and offer workable output with a limited airflow input.

5. Conclusions

This paper investigates the airflow energy harvester using the push–pull diamagnetic levitation structure. Instead of using the three-notch rotor, a four-notch rotor was introduced into the energy harvester, and a corresponding comparison study was carried out to verify the improvement. The centrosymmetric rotors are required to eliminate the periodic offset. Simulations in COMSOL Multiphysics 5.6 of the output performance of multiple floating rotors, have determined that the four-notch rotor has the best output performance.

After the improvement, the levitation characteristics of the four-notch rotor were changed due to its reduced magnetic induction strength and mass.

The axial magnetic spring stiffness increased by an overall average of 9.666%, and the maximum monostable levitation space increased by an average of 1.67%. However, there is a reduction in horizontal response force of approximately 3.97%, which does not affect horizontal stability due to the elimination of the floating rotor horizontal offset. An experimental platform was set up to verify the output and levitation characteristics. According to the experimental results, when the airflow is 3000 sccm, the peak voltage of the four-notch rotor can reach 2.709 V, and the maximum speed can reach 21,360 rpm. The peak voltage, maximum rotation speed, output power, energy conversion factor, and energy conversion rate of the four-notch rotor are respectively increased by 40.80%, 5.99%, 41.40%, 32.68%, and 41.39%, compared to the three-notch rotor. At the same time, the variation in levitation characteristics is consistent with the simulation results. The use of the four-notch rotor also substantially improves output performance, particularly by increasing the energy conversion rate from 41.12% to 58.14%, while increasing the maximum monostable levitation space and axial magnetic spring stiffness, demonstrating that this is a proven improvement solution.

Author Contributions: Conceptualization, L.Z., Y.S. and H.S.; methodology, L.Z., Y.S. and H.S.; software, L.Z., H.S., Y.S. and J.Z.; validation, L.Z., H.S. and Y.S.; formal analysis, L.Z., H.S., J.Z. and Y.S.; resources, H.S. and Y.S.; data curation, L.Z.; writing—original draft preparation, L.Z.; writing—review and editing, Y.S., D.L. and K.C.A.; visualization, L.Z.; supervision, Y.S.; project administration, Y.S.; funding acquisition, Y.S. All authors have read and agreed to the published version of the manuscript.

Funding: This study is supported by the National Natural Science Foundation of China under grant no. U1904169, China Scholarship Council under grant no. 202107045007 and Henan Province Science and Technology Research Project under grant no. 232102220005.

Data Availability Statement: The data presented in this study are available on request from the corresponding author.

Acknowledgments: We would like to extend our thanks to National Natural Science Foundation of China, China Scholarship Council and Henan Province Science and Technology Research Project for the funding of this research.

Conflicts of Interest: The authors declare no conflict of interest.

References

1. Abdulkarem, M.; Samsudin, K.; Rokhani, F.Z.; Rasid, M.F. Wireless sensor network for structural health monitoring: A contemporary review of technologies, challenges, and future direction. *Struct. Health Monit.* **2020**, *19*, 693–735. [CrossRef]
2. Karimzadehkhouei, M.; Ali, B.; Ghourichaei, M.J.; Alaca, B.E. Silicon Nanowires Driving Miniaturization of Microelectromechanical Systems Physical Sensors: A Review. *Adv. Eng. Mater.* **2023**, *25*, 2300007. [CrossRef]
3. Jeyaraj, R.; Balasubramaniam, A.; Kumara, M.A.A.; Guizani, N.; Paul, A. Resource Management in Cloud and Cloud-Influenced Technologies for Internet of Things Applications. *ACM Comput. Surv.* **2023**, *55*, 1–37. [CrossRef]
4. Jiang, B.; Zhu, Y.; Zhu, J.; Wei, X.; Dai, H. An adaptive capacity estimation approach for lithium-ion battery using 10-min relaxation voltage within high state of charge range. *Energy* **2023**, *263*, 125802. [CrossRef]
5. Boukoberine, M.N.; Zhou, Z.; Benbouzid, M. A critical review on unmanned aerial vehicles power supply and energy management: Solutions, strategies, and prospects. *Appl. Energy* **2019**, *255*, 113823. [CrossRef]
6. Hernandez-Benitez, A.; Balam, A.; Vazquez-Castillo, J.; Estrada-Lopez, J.J.; Quijano-Cetina, R.; Bassam, A.; Aviles, F.; Castillo-Atoche, A. An Ultra-Low-Power Strain Sensing Node for Long-Range Wireless Networks in Carbon Nanotube-Based Materials. *IEEE Sens. J.* **2022**, *22*, 9778–9786. [CrossRef]
7. Kato, M. Numerical Simulation on Electromagnetic Energy Harvester Oscillated by Speed Ripple of AC Motors. *Energies* **2023**, *16*, 940. [CrossRef]
8. Holm, P.; Imbaquingo, C.; Mann, B.P.; Bjork, R. High power electromagnetic vibration harvesting using a magnetic dumbbell structure. *J. Sound Vibr.* **2023**, *546*, 117446. [CrossRef]
9. Yang, X.; Zheng, H.; Shao, J.; Zhang, Y.; Chen, Y. Output Characteristics of an Electromagnetic-Triboelectric Hybrid Energy Harvester Based on Magnetic Liquid. *ACS Appl. Electron. Mater.* **2023**, *5*, 775–783. [CrossRef]

10. Shan, C.; Li, K.; Cheng, Y.; Hu, C. Harvesting Environment Mechanical Energy by Direct Current Triboelectric Nanogenerators. *Nano-Micro Lett.* **2023**, *15*, 127. [CrossRef]
11. Li, M.; Luo, A.; Luo, W.; Liu, X.; Wang, F. Electrostatic Vibration Energy Harvester with a Self-Rechargeable Electret. *IEEE Electron Device Lett.* **2023**, *44*, 540–543. [CrossRef]
12. Zhai, L.; Gao, L.; Wang, Z.; Dai, K.; Wu, S.; Mu, X. An Energy Harvester Coupled with a Triboelectric Mechanism and Electrostatic Mechanism for Biomechanical Energy Harvesting. *Nanomaterials* **2022**, *12*, 933. [CrossRef] [PubMed]
13. Lin, Z.; Li, H.; Lv, S.; Zhang, B.; Wu, Z.; Yang, J. Magnetic Force-Assisted Nonlinear Three-Dimensional Wideband Energy Harvester Using Magnetostrictive/Piezoelectric Composite Transducers. *Micromachines* **2022**, *13*, 1633. [CrossRef]
14. Yamaura, S.; Nakajima, T.; Kamata, Y.; Sasaki, T.; Sekiguchi, T. Production of vibration energy harvester with impact-sliding structure using magnetostrictive Fe-Co-V alloy rod. *J. Magn. Magn. Mater.* **2020**, *514*, 167260. [CrossRef]
15. Jeon, B.; Yoon, C.S.; Yoon, W. Experimental Study on Zinc Oxide Thin Film-Based Thermoelectric Energy Harvester Under Plane-Vertical Temperature Gradients. *IEEE Sens. J.* **2021**, *21*, 27298–27307. [CrossRef]
16. Zhu, P.; Luo, X.; Lin, X.; Qiu, Z.; Chen, R.; Wang, X.; Wang, Y.; Deng, Y.; Mao, Y. A self-healable, recyclable, and flexible thermoelectric device for wearable energy harvesting and personal thermal management. *Energy Conv. Manag.* **2023**, *285*, 117017. [CrossRef]
17. Ma, T.; Ding, Y.; Wu, X.; Chen, N.; Yin, M. Research on piezoelectric vibration energy harvester with variable section circular beam. *J. Low Freq. Noise Vib. Act. Control* **2020**, *40*, 753–771. [CrossRef]
18. Su, Y.; Liu, Y.; Li, W.; Xiao, X.; Chen, C.; Lu, H.; Yuan, Z.; Tai, H.; Jiang, Y.; Zou, J. Sensing-transducing coupled piezoelectric textiles for self-powered humidity detection and wearable biomonitoring. *Mater. Horizons* **2023**, *10*, 842–851. [CrossRef]
19. Hsu, T.; Wu, H.; Tsai, D.; Wei, C. Photovoltaic Energy Harvester with Fractional Open-Circuit Voltage Based Maximum Power Point Tracking Circuit. *IEEE Trans. Circuits Syst. II-Express Briefs* **2019**, *66*, 257–261. [CrossRef]
20. Sakthivel, K.; Krishnasamy, R.; Balasubramanian, K.; Krishnakumar, V.; Ganesan, M. Averaged state space modeling and the applicability of the series Compensated Buck-Boost converter for harvesting solar Photo Voltaic energy. *Sustain. Energy Technol. Assess.* **2022**, *53*, 102611. [CrossRef]
21. Xin, M.; Jiang, X.; Xu, C.; Yang, J.; Lu, C. Two-Dimensional Omnidirectional Wind Energy Harvester with a Cylindrical Piezoelectric Composite Cantilever. *Micromachines* **2023**, *14*, 127. [CrossRef]
22. Wang, S.; Liao, W.; Zhang, Z.; Liao, Y.; Yan, M.; Kan, J. Development of a novel non-contact piezoelectric wind energy harvester excited by vortex-induced vibration. *Energy Conv. Manag.* **2021**, *235*, 113980. [CrossRef]
23. Wang, L.; Zhu, D. A Flapping Airflow Energy Harvester with Flexible Wing Sections. In Proceedings of the 19th International Conference on Micro and Nanotechnology for Power Generation and Energy Conversion Applications (Power MEMS), Krakow, Poland, 2–6 December 2019.
24. Cansiz, A.; Hull, J. Stable load-carrying and rotational loss characteristics of diamagnetic bearings. *IEEE Trans. Magn.* **2004**, *40*, 1636–1641. [CrossRef]
25. Shao, H.; Li, X.; Cheng, S.; Aw, K.C.; Su, Y. Comparative Analysis and Experiment Verification of Diamagnetically Stable Levitation Structure. *IEEE Sens. J.* **2023**, *23*, 11469–11481. [CrossRef]
26. Shao, H.; Bian, F.; Zhang, L.; Zhang, J.; Aw, K.C.; Su, Y. Airflow Energy Harvester based on Diamagnetic Levitation Structure with Double Stabilizing Magnets. *IEEE Sens. J.* **2023**, *23*, 13942–13956. [CrossRef]
27. Simon, M.D.; Geim, A.K. Diamagnetic levitation: Flying frogs and floating magnets (invited). *J. Appl. Phys.* **2000**, *87*, 6200–6204. [CrossRef]
28. Cheng, S.; Li, X.; Wang, Y.; Su, Y. Levitation Characteristics Analysis of a Diamagnetically Stabilized Levitation Structure. *Micromachines* **2021**, *12*, 982. [CrossRef] [PubMed]

 *micromachines*

Article

CMOS Radio Frequency Energy Harvester (RFEH) with Fully On-Chip Tunable Voltage-Booster for Wideband Sensitivity Enhancement

Yizhi Li [1], Jagadheswaran Rajendran [1,*], Selvakumar Mariappan [1], Arvind Singh Rawat [1], Sofiyah Sal Hamid [1], Narendra Kumar [2], Masuri Othman [3] and Arokia Nathan [4]

[1] Collaborative Microelectronics Design Excellence Centre (CEDEC), Universiti Sains Malaysia, Bayan Lepas 11900, Malaysia
[2] Department of Electrical Engineering, Faculty of Engineering, University of Malaya, Kuala Lumpur 50603, Malaysia
[3] Institute of Microengineering and Nanoelectronics, National University of Malaysia, Bangi 43600, Malaysia
[4] Darwin College, Cambridge University, Cambridge CB3 9EU, UK
* Correspondence: jaga.rajendran@usm.my

Abstract: Radio frequency energy harvesting (RFEH) is one form of renewable energy harvesting currently seeing widespread popularity because many wireless electronic devices can coordinate their communications via RFEH, especially in CMOS technology. For RFEH, the sensitivity of detecting low-power ambient RF signals is the utmost priority. The voltage boosting mechanisms at the input of the RFEH are typically applied to enhance its sensitivity. However, the bandwidth in which its sensitivity is maintained is very poor. This work implements a tunable voltage boosting (TVB) mechanism fully on-chip in a 3-stage cross-coupled differential drive rectifier (CCDD). The TVB is designed with an interleaved transformer architecture where the primary winding is implemented to the rectifier, while the secondary winding is connected to a MOSFET switch that tunes the inductance of the network. The TVB enables the sensitivity of the rectifier to be maintained at 1V DC output voltage with a minimum deviation of −2 dBm across a wide bandwidth of 3 to 6 GHz of 5G New Radio frequency (5GNR) bands. A DC output voltage of 1 V and a peak PCE of 83% at 3 GHz for −23 dBm input power are achieved. A PCE of more than 50% can be maintained at the sensitivity point of 1 V with the aid of TVB. The proposed CCDD-TVB mechanism enables the CMOS RFEH to be operated for wideband applications with optimum sensitivity, DC output voltage, and efficiency.

Keywords: radio frequency energy harvester; CMOS; voltage boosting; wideband; CCDD; 5GNR

Citation: Li, Y.; Rajendran, J.; Mariappan, S.; Rawat, A.S.; Sal Hamid, S.; Kumar, N.; Othman, M.; Nathan, A. CMOS Radio Frequency Energy Harvester (RFEH) with Fully On-Chip Tunable Voltage-Booster for Wideband Sensitivity Enhancement. *Micromachines* **2023**, *14*, 392. https://doi.org/10.3390/mi14020392

Academic Editors: Qiongfeng Shi and Huicong Liu

Received: 28 November 2022
Revised: 17 January 2023
Accepted: 18 January 2023
Published: 4 February 2023

1. Introduction

Since the 3rd Generation Partnership Project (3GPP) standardized the first New Radio (NR) version (Release 15) in mid-2018, the evolution of the fifth-generation (5G) new radio (NR) has been rapid. 5GNR technology was designed to operate in two distinct bands: FR1 (410 to 7125 MHz) and FR2 (24,250 to 52,600 MHz). Despite operating into the 7 GHz band, FR1 is commonly referred to as the "Sub-6 GHz" band. The leading carriers compete to provide various commercial services over 5G networks. In the future, it is expected that over 6.5 million 5G base stations will be installed, allowing over 58% of the world's population to access services via over 100 billion 5G connections [1]. Following 5G's rapid development, many commercialization use cases will push the 5G network to improve performance and expand capabilities continuously.

It is no secret that the worldwide demand for electronic devices has been rising, which means that the global use of electrical energy has been growing alongside it. According to studies cited in [2], by 2030, communication technologies might account for as much as 51% of the world's total electricity consumption. The study also indicates that by 2030, the power used by communication technologies might account for up to 23% of worldwide greenhouse gas emissions. As a result, reducing the energy needed to run all these gadgets connected to the internet is an absolute necessity. Furthermore, these connected devices are powered by batteries, contributing to increased waste and environmental pollution. Since the introduction of the first commercial lithium-ion (Li-Ion) batteries in 1991, the number of portable electronic device products that use Li-Ion batteries has grown dramatically [3].

Since the battery life of connected electronic devices can be vastly improved by reducing the frequency with which they need to be charged, it is crucial to lower their power consumption. Several energy harvesting works have been successfully implemented recently, including the triboelectric nano-generators (TENG) devices [4,5]. TENG devices are commonly utilized for biomechanical energy harvesting applications. In [4], a nanocomposite that acts as a positive triboelectric layer was successfully used in the fabrication of the TENG device, which converts waste mechanical energy into valuable electrical energy. The TENG device consumes an area of $4\,\text{cm}^2$ while capable of delivering an output voltage of 35 V as well as 130 nA output current at 100 MΩ load. Moreover, in [5], a ZIF-8 HG-Kapton TENG device with dual-mode operation was proposed. A triple-unit TENG is constructed using additive manufacturing, producing an output voltage of 150 V and a current of 4.95 µA. The single-unit mode is implemented to detect a robotic system's right and left tilting motion. Meanwhile, the triple-unit mode is utilized as a sustainable power source to power up low-power electronics. However, the TENG devices mainly focus on harvesting the wasted mechanical motions, which is not present in on-chip microelectronics integrated circuits.

Radio frequency energy harvesting (RFEH) is a particularly prevalent type of renewable energy harvesting at the moment [6–11] because RFEH can be unified into a single wireless communication system used by all interconnected electronic devices. RFEH is widely studied and implemented in integrated circuits to improve the system's efficiency and charge the batteries. It aids in decreasing the need for large battery packs in electronic gadgets and the amount of time between charges. The RFEH system consists of an antenna, an input impedance matching network, a rectifier, a voltage multiplier, and an energy storage component or load.

The antenna in the RFEH system detects or receives the transmitted RF wave propagating through the transmission medium. Its impedance-matching network is designed to provide maximum power transfer for the RFEH system. The impedance matching network employs low-loss or high Q-factor inductors or transformers. The rectifier converts the received alternating current (AC) RF signal to direct current (DC) power. However, the rectifier's DC output voltage is too low to power a wireless device. As a result, a voltage multiplier is used to increase the DC voltage level to the level required by wireless devices. Finally, the harvested energy is stored in energy storage devices such as rechargeable batteries and super-capacitors or directly applied to the load. Figure 1 depicts the block diagram of an RFEH system [12].

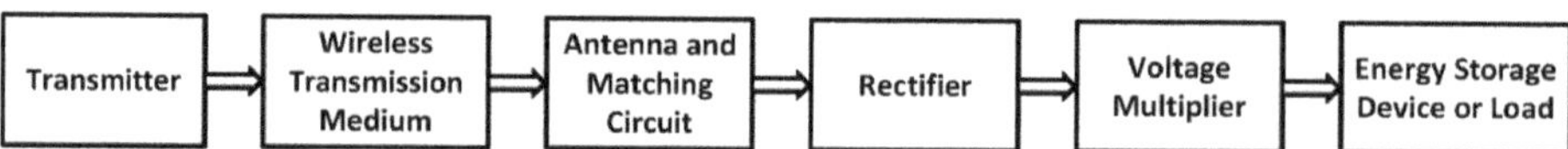

Figure 1. Block diagram of an RFEH system.

A reconfigurable RFEH with dual-path rectifiers and an adaptive control circuit (ACC) is presented in [13]. The dual-path rectifiers are configured with series and parallel paths for low-power and high-power operations. This method maintains a high-power PCE over a wide range of input power. The rectifiers use internal threshold voltage cancellation (IVC) to compensate the transistors' V_{th} effectively. The ACC includes a comparator, an inverter, and three switches that activate series or parallel paths based on the RF input power range. An off-chip impedance matching network is used to match the rectifier's input impedance to the source of the RF input power. For a 200 kΩ load at 0.902 GHz, the reconfigurable

RFEH achieved a peak power conversion efficiency (PCE) of 33% and a DC output voltage of 3.23 V with -8 dBm RF input power. It has a sensitivity of -20.2 dBm input power and outputs 1 V DC voltage for a 1 MΩ load.

Furthermore, a 3-stage Cross-Coupled Differential-Drive (CCDD) rectifier with a broad PCE dynamic range was proposed in [14]. The proposed RFEH uses a self-body biasing approach to lower the V_{th} and diode-tied transistors to lower the reverse leakage current. Both shared-capacitor-coupling (SCC) and individual-capacitor-coupling (ICC) input-capacitor topologies are designed independently for the CCDD rectifier. Two separate capacitors are used to separate the DC power supply from the RF signal route. When used in conjunction with one another, these capacitors bias the gate of the transistors while one store charge. The SCC configuration is achieved when the NMOS and PMOS transistor gates are linked to the same coupling capacitor at the input. In contrast, the gate biasing of each NMOS and PMOS transistor is kept independent by physically separating the coupling capacitor in the ICC arrangement. Input power of -18.4 dBm at 0.9 GHz resulted in a peak PCE of 83.7% for the SCC configuration. At the same frequency, the ICC configuration's peak PCE was 80.3% when fed with an input power of -17 dBm.

Moreover, an RFEH based on an improved Dickson charge pump with an output capacitor to reduce the load capacitance during the positive half cycle is proposed in [15]. The rectifiers addressed here are modified output capacitor Dickson and differential load Dickson. The Dickson charge pump scheme is used in both rectifiers, which consists of diode-connected transistors, charge storage coupling capacitors, load capacitors for output ripple reduction, and resistive load. The first design incorporates a 3-stage voltage multiplier with a modified output capacitor loop. In the meantime, the second design employs a 2-stage voltage multiplier with a differential load. The output capacitor helps increase the DC output voltage, allowing for appropriate load resistors. The differential load, on the other hand, helps to improve the rectifier's sensitivity to operate for low RF input signals at 0.953 GHz. Under -12.5 dBm input RF power, the rectifier with a modified output capacitor achieved a peak PCE of 84.4% and 1 V DC output voltage. Meanwhile, under -15 dBm input power, the rectifier with differential capacitor load achieved a peak PCE of 56.2%.

As observed from recent on-chip RFEH works, it is evident that most of them are operating in narrowband applications, and a wideband energy harvesting solution is still a challenge, unlike in other RF circuits, such as power amplifiers or voltage-controlled oscillators, which achieve wideband [16,17]. Thus, in this study, a 3-stage CCDD rectifier is equipped with an on-chip tunable voltage boosting mechanism that is entirely adjustable for wideband sensitivity tuning. The tunable voltage booster is constructed with an interleaved transformer architecture, with the primary winding connected to the rectifier while the secondary winding is connected to a MOSFET switch that tunes the network's inductance. Adjusting the voltage booster enables the CMOS RFEH's sensitivity to be maintained at 1V DC output voltage throughout a bandwidth of 3 to 6 GHz of 5GNR frequency bands. The suggested technique allows the CMOS RFEH to operate with optimal sensitivity, DC output voltage, and PCE for wideband applications.

2. Rectifier Design and Operation

The rectifier designed adopts the cross-coupled differential-drive (CCDD) architecture, as depicted in Figure 2. Referring to Figure 2, the CCDD rectifier comprises two NMOS transistors (M_{N1} and M_{N2}) and two PMOS transistors (M_{P1} and M_{P2}) connected in a cross-coupled configuration. The transistors in the rectifier are operating in the subthreshold region, and the drain-current I_D is given as [18]:

$$I_D = I_S \cdot e^{\frac{V_{GS} - V_{TH}}{nV_T}} \left(1 - e^{-\frac{V_{DS}}{V_T}} \right) \tag{1}$$

where I_S is the zero-bias current of the device, V_{GS} is the gate-source voltage of the transistor, V_{DS} is the drain-source voltage of the transistor, V_{TH} is the threshold voltage of the transistor, n is the subthreshold slope, and V_T is the thermal voltage.

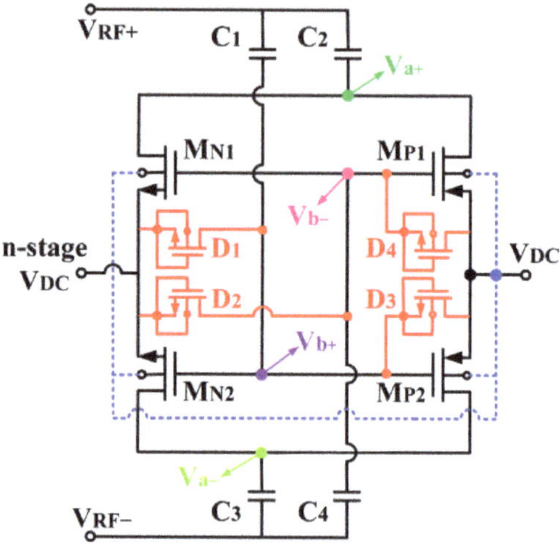

Figure 2. Schematic of the designed CCDD rectifier.

M_{N1} and M_{N2} adopt a self-body-biasing mechanism in which their bulk terminals are connected to the DC output voltage (V_{DC}). The self-body-biasing mechanism reduces the threshold gate voltage (V_{TH}) of M_{N1} and M_{N2}. Figure 3 compares drain current (I_D) with and without bulk biasing. It can be observed that the V_{TH} of the transistor can be mitigated to 440 mV from 540 mV after applying the bias to the bulk. The transistor's V_{TH} contributes to voltage drop, limiting the maximum V_{DC} and the PCE that can be achieved. Therefore, it is essential to lower the transistor's V_{TH} for the rectifier. Furthermore, to mitigate the reverse-leakage current, diode-connected MOS transistors (D_1–D_4) are integrated between the gate and the source of each transistor, as depicted in Figure 2. The diodes also aid in raising the forward conduction of the rectifier.

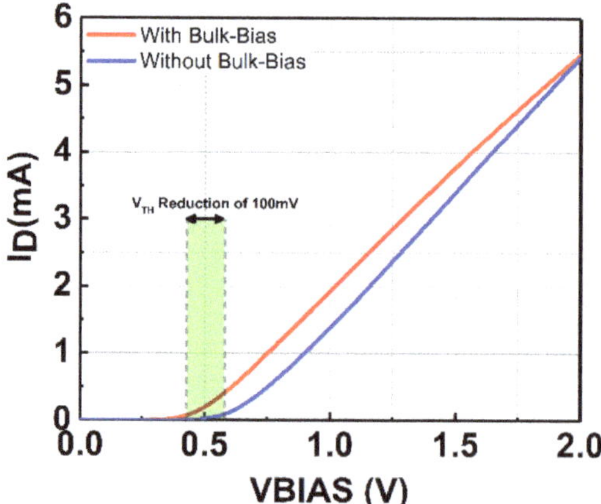

Figure 3. Drain current (I_D) comparison without and with bulk-biasing. 100 mV V_{TH} reduction is achieved with a bulk-biasing mechanism.

Assuming a steady-state operation, when the RF input power of the rectifier is low and for the first half-period of the input RF signal ($V_{RF+} > V_{RF-}$), M_{P1} and M_{N2} are turned on to perform rectification while M_{P2} and M_{N1} are turned off. For the second half-period of the input RF signal, M_{P2} and M_{N1} perform the rectification function while M_{P1} and M_{N2} are turned off. During the first and second half-periods, the bulk of M_{N2} and M_{N1} are, respectively, biased by the V_{DC} from the rectifier in which it realizes the self-body-biasing mechanism. The self-body-biasing is not implemented on M_{P1} and M_{P2} since only the positive DC voltage is produced in the rectifier circuitry. At the low RF input signal, the reverse leakage current is minimal as the resistive loss mainly contributes to the energy loss during forward conduction. This is due to the low DC biasing voltage applied at the transistor's bulk, which is not significant for V_{TH} reduction. A voltage of more than 500 mV is necessary to overcome the forward voltage of the diode-configured transistors utilized in the rectifier circuit. This is shown in Figure 3, where the I-V curve of the diode configured MOS with W/L of 12.0 μm/0.13 μm begins to conduct current beyond 540 mV. Thus, D_1–D_4 is turned off when the RF input signal is low.

On the other hand, when the RF input signal is large enough to overcome the forward voltage of D_1–D_4, it starts to operate. At M_{N1} and M_{N2}, D_1 and D_2 are in reverse biased condition when n-stage $V_{DC} > V_{RF\pm}$. n-stage V_{DC} is the DC output voltage from the previous stage of the rectifier. The potential difference between the source and gate creates an extra current path for the forward current to flow to the system's drain, increasing the PCE. Suppose the voltage $V_{DC} > V_{RF\pm}$, the D_1, and D_2 are in forward biased condition. The diodes act as a large resistor that blocks the reverse-leakage current flowing from n-stage V_{DC} into $V_{RF\pm}$. Moreover, when $V_{DC} < V_{RF\pm}$, D_3, and D_4, which are connected at the M_{P2} and M_{P1}, respectively, are in reverse bias. D_3 and D_4 emulate a huge resistor with a potential of $-V_D$ between the gate and source of the transistors to block the reverse leakage current from flowing back to the source. Suppose the voltage $V_{DC} > V_{RF\pm}$, D_3, and D_4 are in forward bias, where it increases the forward current by pushing further charge to the output, increasing the total output charge and thus enhancing V_{DC}, sensitivity, and PCE of the rectifier.

As depicted in Figure 2, the proposed CCDD rectifier is implemented with a mutual capacitor coupling configuration as the gates of the NMOS and PMOS are connected to the same coupling capacitor at the input. The capacitors pair (C_2 and C_3) decouple from the gates of the transistors to isolate the RF signal path from the power line with the addition of C_1 and C_4. C_2 and C_3 serve as charge-storing capacitors only, while C_1 and C_4 focus solely on biasing the gate of the transistors without being impacted by the DC offset contributed by C_2 and C_3. Figure 4 depicts the input RF voltage waveforms for the proposed CCDD rectifier.

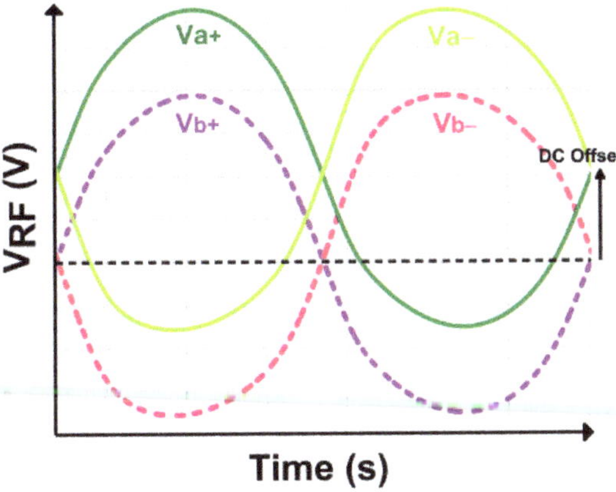

Figure 4. Terminal RF voltage waveform of the designed CCDD rectifier.

When implemented in ambient RFEH systems, cascading more stages typically allows for a higher V_{DC} output. Albeit, as a result of its proportionality to RF input power (P_{RFIN}) and PCE for a fixed output load (R_L), the power equation in (2) imposes a theoretical limit on the possible value of V_{DC}. The PCE of the rectifier is defined as in (3). As demonstrated in Figure 5, extending the rectifier stages beyond necessary would be superfluous because V_{DC} would not increase considerably. By referring to Figure 5, it can be deduced that the V_{DC} is not rising significantly beyond 3 stages and begins to drop as the stages increase. Therefore, a 3-stage rectifier is considered essential, and its configuration is depicted in Figure 6. Meanwhile, Figure 7 shows the rectification response of the 3-stage CCDD rectifier at 3 GHz.

$$V_{DC} = \sqrt{\eta \cdot P_{RFIN} \cdot R_L} \tag{2}$$

$$\eta = \frac{P_{DC}}{P_{RFIN}} = \frac{V_{DC}^2}{R_L \cdot P_{RFIN}} \times 100\% \tag{3}$$

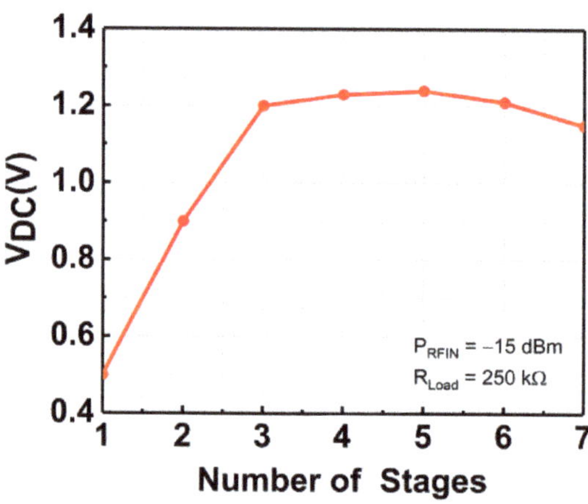

Figure 5. V_{DC} versus rectifier stages.

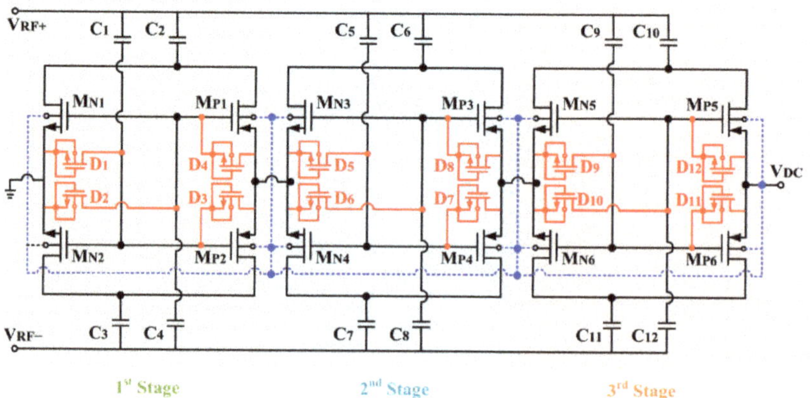

Figure 6. Schematic of the 3-stage CCDD rectifier.

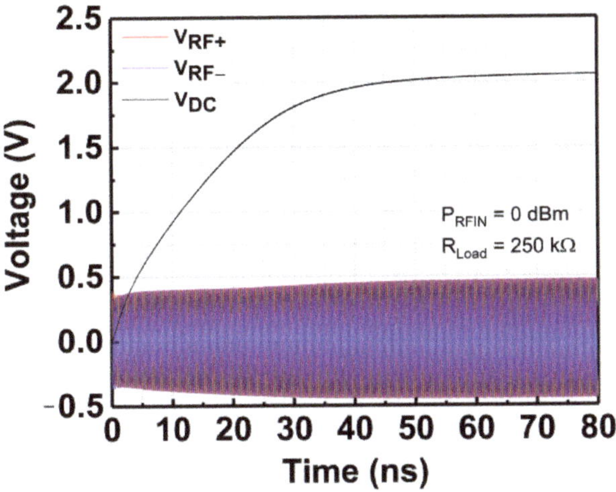

Figure 7. Rectification response of the 3-stage CCDD rectifier at 3 GHz.

3. Tunable Voltage Booster (TVB) Design and Operation

The tunable voltage booster (TVB) consists of a primary conductor and a secondary conductor interleaved between each other, as illustrated in Figure 8. The primary conductor is the main inductor connected to the input of the rectifier. In contrast, the secondary conductor is the auxiliary conductor used to tune the inductance of the primary conductor through magnetic coupling. A transistor switch is connected in parallel to the secondary conductor, which opens and shorts the conductor when the gate voltage is applied. A coupling coefficient, k, magnetically couples the primary conductor and secondary conductor. When the ambient RF input signal is applied to the primary conductor, the changing magnetic field in the primary conductor induces an opposite current flow which translates to an opposing magnetic field on the secondary conductor.

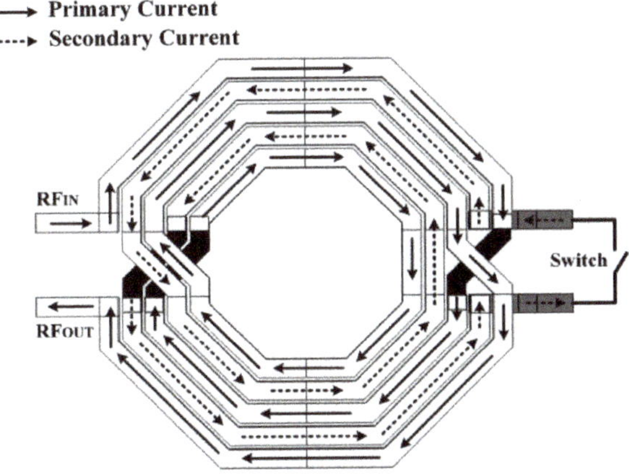

Figure 8. Tunable voltage booster (TVB) architecture.

The current flow in the secondary conductor is gradually shorted by varying the gate voltage of the transistor switch. When the transistor switch is entirely off, the secondary coil is open. Thus, no current flow occurs, and the inductance value in the primary conductor remains the default. When the transistor switch is gradually turned on via the applied gate voltage, the secondary conductor becomes shorted in, where the current flow opposes a change of the flux. The opposing magnetic field in the secondary conductor cancels out the magnetic field created in the primary conductor. As a result, the decreasing magnetic field reduces the inductance of the primary conductor, which realizes the variability of the inductance value.

Figure 9 shows the schematic of the TVB configuration applied in the CCDD rectifier. The ambient RF input signal is applied at the primary conductor, L_p. The secondary conductor, L_s, is connected to the drain and source of an NMOS transistor, acting as a switch. A DC voltage (V_T) is applied at the gate of the NMOS transistor, where a resistor, R_1, is connected in series to provide isolation between the AC signal path and the DC signal path. The secondary conductor is open when the transistor is turned off ($V_T = 0$ V). When the gate voltage is gradually increased ($V_T > 0$ V), the secondary conductor is shorted, which induces an opposite current flow that opposes the magnetic field in the primary conductor, as aforementioned. The current flow paths on each conductor are illustrated in Figure 8. The opposite current flow in the secondary reduces the inductance value in the primary conductor due to their magnetic coupling. The following brief analysis, with the aid of Figure 10, shows the relation of primary inductance when the secondary coil is shorted. Referring to Figure 10:

$$V_1 = j\omega L_P I_1 + j\omega M I_2 \tag{4}$$

$$V_2 = j\omega M I_1 + j\omega L_s I_2 \tag{5}$$

when L_s is shorted,

$$V_2 = 0 \tag{6}$$

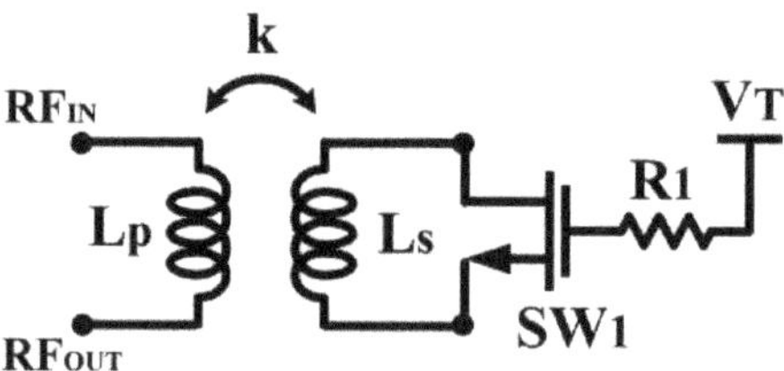

Figure 9. Schematic of the TVB.

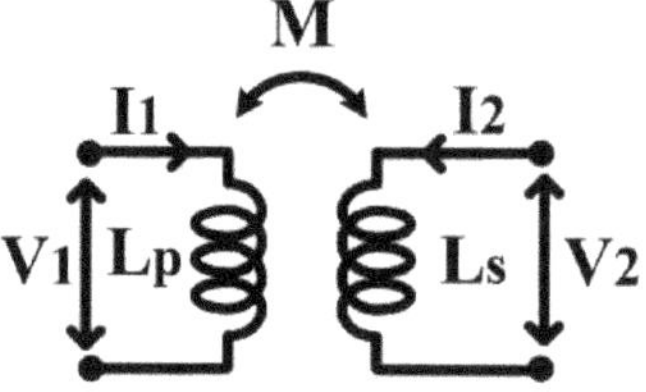

Figure 10. Simple analysis schematic for the TVB operation.

Thus,

$$I_2 = -\frac{M}{L_s} I_1 \tag{7}$$

Substituting (7) into (4),

$$V_1 = j\omega L_P I_1 + j\omega M \left(-\frac{M}{L_s} I_1\right) \tag{8}$$

$$V_1 = j\omega L_P I_1 + j\omega M \left(-\frac{M}{L_s} I_1\right) \tag{9}$$

Since $M = k\sqrt{L_P L_s}$:

$$M^2 = k^2 L_P L_s \tag{10}$$

Substituting (10) into (9),

$$V_1 = j\omega L_P\left(1 - k^2\right)I_1 \tag{11}$$

Thus, the inductance of L_P when the switch is turned on is:

$$L_{P,SW=ON} = L_P\left(1 - k^2\right) \tag{12}$$

It can be deduced from (12) that the inductance of L_P is reduced when the switch is turned on at the secondary coil.

The TVB is designed and characterized according to the impedance values needed for the rectifier to achieve wideband sensitivity from 3 to 6 GHz. Figure 11a delineates the simulated inductance value variations of the TVB when VT is tuned from 0 to 2.25 V. Referring to Figure 11a, it can be observed that the inductance (L) can be varied from 5 nH to 1.6 nH at 3 GHz. It is also observed that the inductor's self-resonance frequency (SRF) is increased to a higher frequency when V_T is tuned. On the other hand, Figure 11b shows the respective simulated Q-factor of the inductor. This tuning property of the voltage booster is integrated with the rectifier to achieve the wideband functionality of the CCDD rectifier.

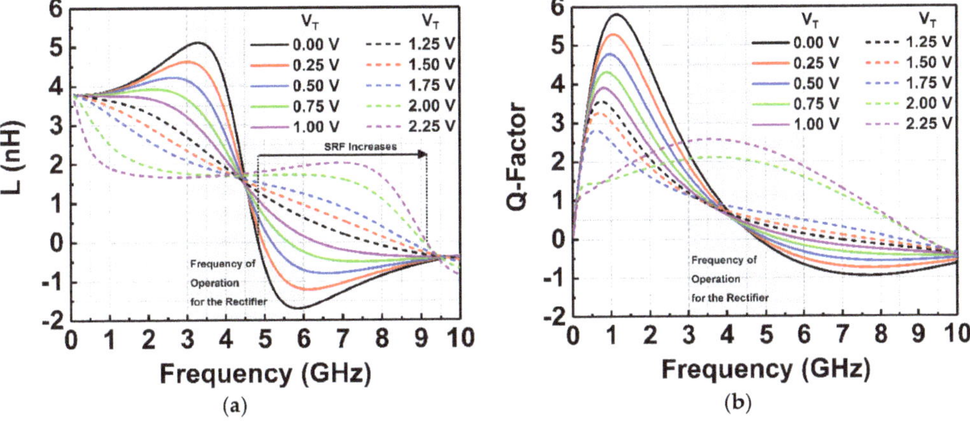

Figure 11. (a) Simulated inductance values of the TVB when V_T is tuned. (b) Simulated Q-factor values of the TVB when V_T is tuned.

4. CCDD-TVB Rectifier Design and Operation

The fully integrated architecture of the CCDD-TVB rectifier is illustrated in Figure 12. The rectifier is constructed in 3 stages to achieve the optimum DC output voltage and PCE. The TVB is employed at both differential inputs of the rectifier to provide the tuning property at the rectifier's input symmetrically. M_{N1} to M_{N6} and M_{P1} to M_{P6} are the NMOS and PMOS transistors utilized in the 3-stage CCDD rectifier. D_1 to D_{12} are the diode-connected transistors employed to suppress the reverse-leakage current throughout the stages. C_2, C_3, C_6, C_7, C_{10}, and C_{11} are the charge storing capacitors, while C_1, C_4, C_5, C_8, C_9, and C_{12} are the gate biasing capacitors. The tunable voltage boosters are L_1 (L_{1P} and L_{1S}), and L_2 (L_{2P} and L_{2S}) applied at the positive (V_{RF+}) and negative (V_{RF-}) inputs of the rectifier, respectively. L_{1P} and L_{1S} are the primary and secondary coil of L_1, respectively, while L_{2P} and L_{2S} are the primary and secondary coil of L_2, respectively. SW_1 and SW_2 are the NMOS switches employed at L_{1S} and L_{2S}, respectively, to tune the inductance of L_{1P} and L_{2P}. R_1 and R_2 are the isolation resistors for both TVBs. The gate voltages for SW_1

and SW_2 are applied via V_{T1} and V_{T2}, respectively, to conduct the tuning mechanism of the TVBs. C_L is the capacitor load of the rectifier, and it is tested with a resistor load (R_L) of 250 kΩ.

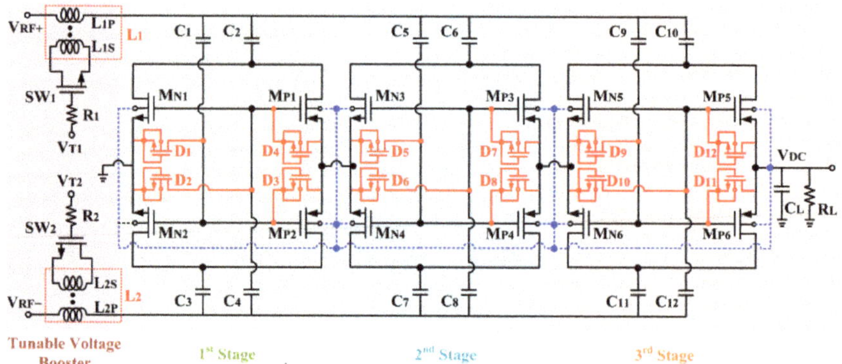

Figure 12. Detailed schematic of the CCDD-TVB rectifier.

Conventionally, a series inductor at the rectifier's input is utilized as the voltage booster (VB). The conventional implementation of the VB has significantly enhanced the V_{DC} of the rectifier. Figure 13 shows the simulated V_{DC} achieved by the rectifier with conventional VB and without VB across the targeted frequency of 3 to 6 GHz. At 3 GHz, the sensitivity of the CCDD rectifier at 1 V V_{DC} is significantly enhanced by -7 dBm (from -17 to -24 dBm). However, when the frequency changes, the sensitivity drastically degrades, reflecting the drawback of the conventional VB. This is due to the change in input impedance after implementing the conventional VB. Observe in Figure 13 that the sensitivity of the rectifier did not vary much without the VB. This is because the input impedance did not change significantly as compared to after VB implementation. It is clearly illustrated in the Smith chart provided in Figure 14. Referring to Figure 14, it can be deduced that the input impedance's location on the Smith chart changes drastically with the conventional VB compared to without the VB. Since the conventional VB does not have a tuning characteristic, it is a limitation for wideband implementation in rectifiers.

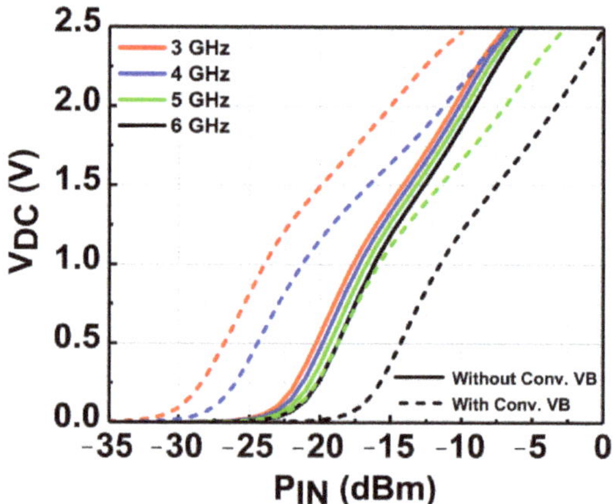

Figure 13. Simulated V_{DC} of the CCDD rectifier with and without the conventional VB.

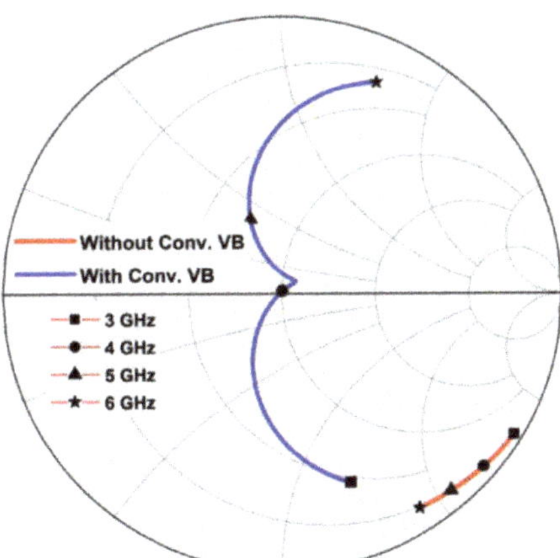

Figure 14. Input impedance's location across frequency with and without conventional VB.

Furthermore, the optimum impedance points for specific parameters can be determined via the load or source pull analysis [19]. Thus, the input impedance for sensitivity across the wideband frequency for the rectifier is identified using the source-pull evaluation. The input impedance for sensitivity at 1 V of less than −20 dBm across frequency is targeted to be achieved. The TVB is designed to fulfill the input impedance needed across the frequency for the targeted sensitivity at 1 V, thus resolving the limitation of the conventional VB, which is narrowband. Figure 15 shows the Smith chart illustrating the input impedance location after tuning the voltage booster. It can be observed that the input impedance's locations for wideband sensitivity can be attained by tuning the TVB. The TVB provides different values of inductances when tuned as compared to conventional VB, in which it can shift the input impedance locations for sensitivity optimization across wideband frequencies.

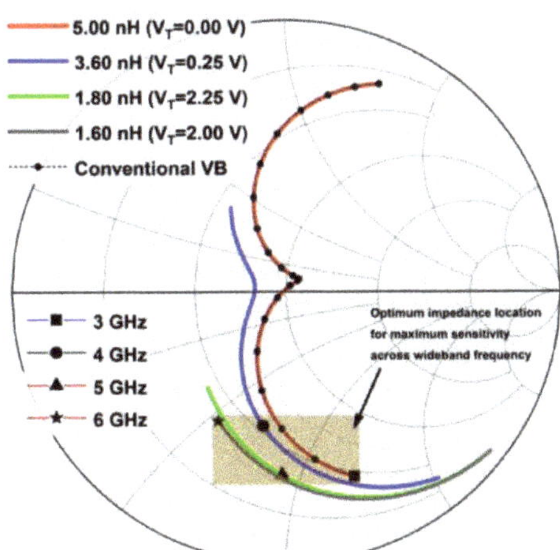

Figure 15. Input impedance's location across frequency after tuning the TVB as compared to conventional VB.

After tuning the TVB, the sensitivity of the rectifier at 1 V can be enhanced across the frequency. Figure 16 depicts the V_{DC} achieved by the rectifier before and after tuning. It can be observed that the sensitivity varies drastically by −13 dBm across frequency before tuning. The sensitivity is enhanced to only -2 dBm variation across frequency after tuning the TVB. It proves the functionality of the tunable voltage in realizing the wideband sensitivity rectifier. At 3 GHz, the best V_T setting is by default 0 V. At 4 GHz; the V_T needed for the optimum performance is 0.25 V. Meanwhile, at 5 and 6 GHz, the V_T settings required are 2.25 V and 2 V, respectively.

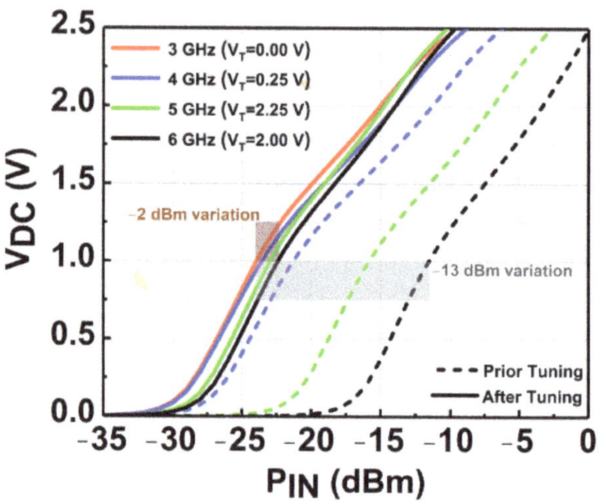

Figure 16. Simulated V_{DC} of the CCDD-TVB rectifier across frequency before and after tuning.

5. Measurement Results

The proposed CCDD-TVB rectifier is fabricated using the technology of CMOS 130 nm and six metal layers. It consumes an area of 1.43 mm^2 on-chip, including the bond pads for measurement. Figure 17 depicts the fabricated chip of the CCDD-TVB rectifier. In addition, Figure 18 illustrates the measurement setup utilized for validating and characterizing the CCDD-TVB rectifier. The parametric analyzer supplies and monitors the DC voltages (V_{T1} and V_{T2}) to the chip. The RF and DC probes are used to probe the connection pads on-chip. Additionally, the multimeter is used to measure the DC output voltage (V_{DC}) of the CCDD-TVB rectifier from the PCB variable load. A 250 kΩ load is used as the optimum load for the system. The signal generator is utilized to supply the RF input power (P_{IN}) to the rectifier and the RF balun to split the signal into a differential signal.

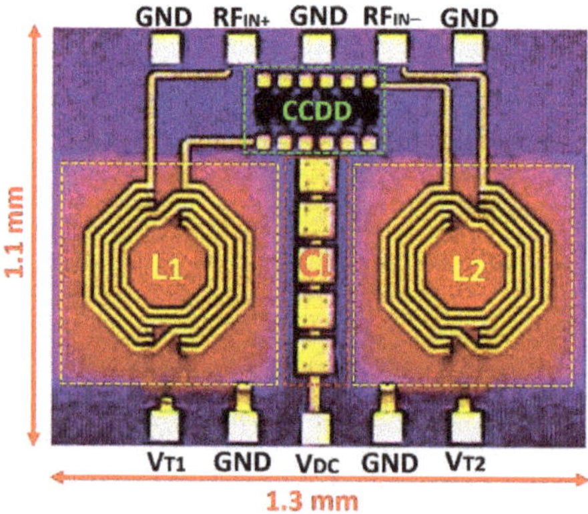

Figure 17. The micro-photograph of the fabricated CMOS 130 nm CCDD-TVB rectifier.

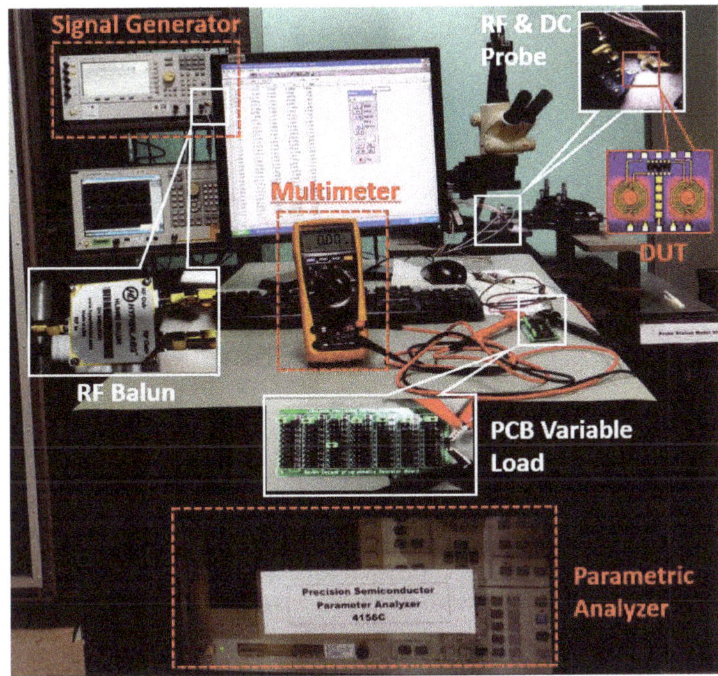

Figure 18. Measurement setup for the validation of CCDD-TVB rectifier.

Figure 19 depicts the measured V_{DC} of the CCDD-TVB rectifier for different load conditions across the P_{IN} at 3 GHz. It can be observed that as the load is increased from 50 kΩ to 250 kΩ, the sensitivity of the rectifier rises as well. The open-load condition is close to the 250 kΩ load, which is required to achieve the highest sensitivity for the rectifier. Thus, 250 kΩ is selected as the optimum load condition required, and the measurement across frequency is conducted. The CCDD-TVB rectifier can deliver a maximum V_{DC} of 2.5 V for P_{IN} of -7 to -8 dBm across the frequency when measured with R_L = 250 kΩ. The measured results do not deviate much from simulated results which reflect the reliability of the design, as illustrated in Figure 20. The measured sensitivity at 1 V obtained is from -21.5 dBm at 6 GHz to -23.5 dBm at 3 GHz, where the variation is -2 dBm across the frequency. Figure 21 delineates the sensitivity of the CCDD-TVB rectifier across the frequency of 3 to 6 GHz. The slight deviation of performance between the simulation and measured results is due to the parasitic effect on-chip and process variations during the fabrication of the rectifier.

Figure 22 shows the respective measured PCE achieved by the CCDD-TVB rectifier across the frequency. A peak PCE of 83% is achieved at 3 GHz for -23 dBm P_{IN}. At 4 GHz, the peak PCE achieved is 73% for -22.5 dBm P_{IN}. At 5 GHz and 6 GHz, the peak PCE is 67% and 57% for -22 dBm and -21.5 dBm, respectively. Referring to Figure 21, the peak PCE is achieved at the lowest sensitivity points across the frequency. It can be observed that the PCE can be maintained at more than 50% across the frequency at its sensitivity point of 1V DC output voltage by tuning the proposed TVB mechanism. Table 1 summarizes the performances of the CCDD-TVB rectifier and compares it to the recent state-of-art rectifiers. It can be observed from the summary that the proposed CCDD-TVB achieves an operating bandwidth of 3 GHz (3 to 6 GHz) as compared to other recent works. This work also realizes the RFEH application for 5GNR bands.

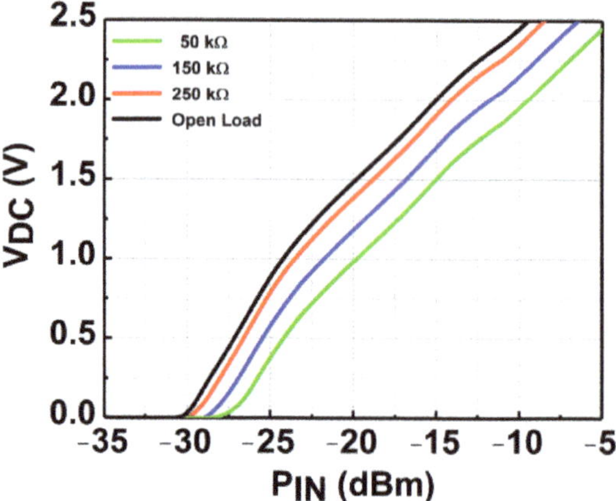

Figure 19. Measured V_{DC} of the CCDD-TVB rectifier at 3 GHz for different load conditions.

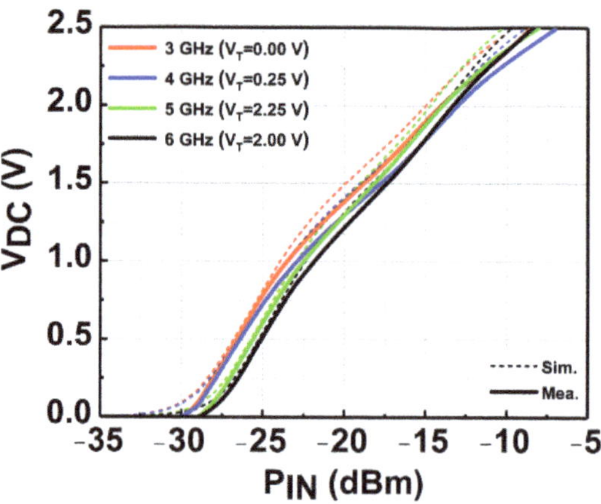

Figure 20. Measured tuned V_{DC} of the CCDD-TVB rectifier across frequency.

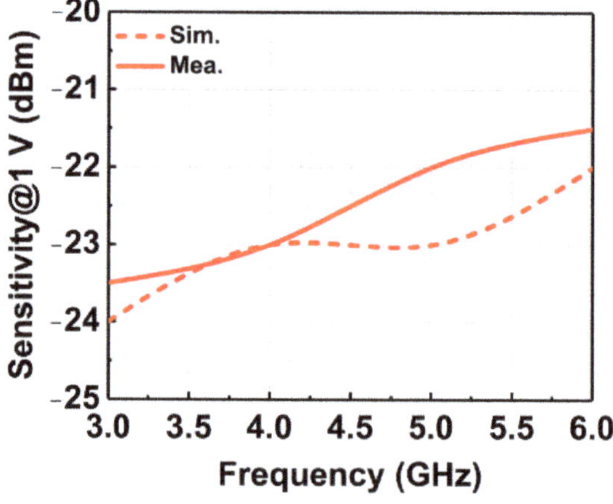

Figure 21. Sensitivity of the CCDD-TVB rectifier across wideband frequency.

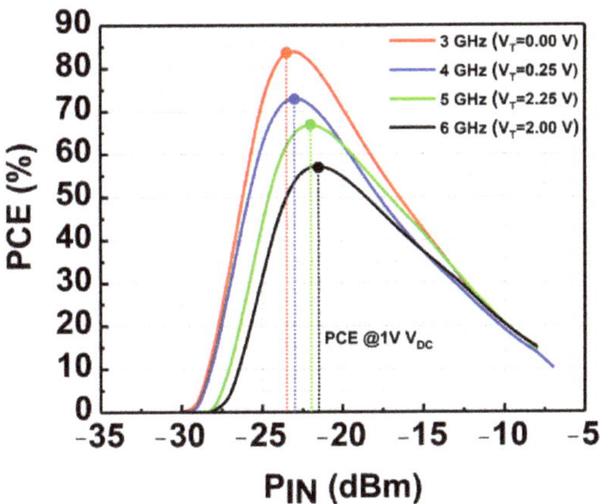

Figure 22. PCE of the CCDD-TVB rectifier across wideband frequency.

Table 1. Comparison of CCDD-TVB rectifier with recent works.

Ref.	Tech.	Rectifier Topology	Freq. (GHz)	Peak PCE (% for Load at Input Level)	VDC$_{out}$ (V)	Sensitivity@VDC (dBm@V)
[13]	180 nm CMOS	1-stage voltage doubler	0.902	33% for 200 kΩ@ −8 dBm	3.23	−20.2@1
[14]	130 nm CMOS	3-stage CCDD	0.9	83.7% for 100 kΩ@ −18.4 dBm	1.1	−19.2@1
[15]	65 nm CMOS	3-stage Dickson	0.953	84.4% for 21.5 kΩ@ −12.5 dBm	1	−12.5@1
[20]	130 nm CMOS	12-stage Dickson	0.915	32% for 1 MΩ@ −15 dBm	3.2	−20.5@1
[21]	180 nm CMOS	Dickson	0.433	30% for 10 kΩ@ −5 dBm	0.5	−9@0.1
[22]	180 nm CMOS	1-stage CCDD	0.1	65% for 100 kΩ@ −18 dBm	1	−18@1
[23]	180 nm CMOS	1-stage CCDD	0.433	65.3% for 50 kΩ@ −15.2 dBm	1	−17@1
[24]	180 nm CMOS	2-stage differential	0.433	74% for 5 kΩ@ −2 dBm	0.8	0@0.8
[25]	180 nm CMOS	5-stage Dickson	0.93/2.63	25.2%/22.5% for 500 kΩ@ −1 dBm	9.5	−16/−15.4@1
[26]	130 nm CMOS	3-stage CCDD Bridge	0.953	72.2% for 10 kΩ@ −1.3 dBm	4.2	−6.3@1
This Work	130 nm CMOS	CCDD + TVB	3–6	83% at 3 GHz for 250 kΩ@ −23 dBm	2.5	−23.5 to −21.5 @1

6. Conclusions

A wideband sensitivity rectifier is designed and validated in this work. The proposed TVB can tune the input impedance of the designed CCDD rectifier, which realizes the wideband property. The CCDD-TVB rectifier can maintain its sensitivity of 1 V V$_{DC}$ with a minimum deviation of −2 dBm input power across a frequency bandwidth of 3 GHz to 6 GHz. The PCE at the sensitivity point is also maintained by more than 40% across the frequency. A peak PCE of 83% is achieved at 3 GHz for P_{IN} of -20 dBm. All the measured results are validated with a 250 kΩ resistor output load for the rectifier. The maximum DC output voltage of 2.5 V is achieved across the bandwidth. The CCDD-TVB architecture realizes the wideband operating rectifier, which is suitable for 5GNR bands that operate from 3 GHz to 6 GHz.

Author Contributions: Conceptualization, Y.L. and J.R.; methodology, Y.L., J.R. and S.M.; software, Y.L. and S.S.H.; formal analysis, Y.L., J.R., S.M., A.S.R. and S.S.H.; validation, Y.L., J.R., S.M., N.K. and M.O.; resources, J.R. and S.M.; data curation, Y.L., J.R., S.M. and A.S.R.; writing—original draft preparation, Y.L., J.R., S.M. and A.S.R.; writing—review and editing, J.R., S.M., N.K., M.O. and A.N.; visualization, Y.L., S.M. and S.S.H.; supervision, J.R., M.O. and A.N.; project administration, J.R.; funding acquisition, J.R. and S.M. All authors have read and agreed to the published version of the manuscript.

Funding: Malaysian Ministry of Higher Education's Fundamental Research Grant Scheme [grant number: FRGS/1/2019/TK04/USM/02/14] and CREST Malaysia [grant PCEDEC/6050415].

Data Availability Statement: Not Applicable.

Acknowledgments: We would like to extend our thanks to Malaysian Ministry of Higher Education's Fundamental Research Grant Scheme [grant number: FRGS/1/2019/TK04/USM/02/14] and CREST Malaysia [grant PCEDEC/6050415] for the funding of this research. Thanks to Collaborative Microelectronics Design Excellence Centre (CEDEC), Universiti Sains Malaysia for supporting the measurement activity of this research. Also, thanks to SilTerra Malaysia Sdn Bhd for the fabrication support.

Conflicts of Interest: The authors declare no conflict of interest.

References

1. Pang, J.; Wang, S.; Tang, Z.; Qin, Y.; Tao, X.; You, X.; Zhu, J. A new 5G radio evolution towards 5G-Advanced. *Sci. China Inf. Sci.* **2022**, *65*, 191301. [CrossRef]
2. Andrae, A.S.G.; Edler, T. On Global Electricity Usage of Communication Technology: Trends to 2030. *Challenges* **2015**, *6*, 117–157. [CrossRef]
3. Mossalia, E.; Picone, N.; Gentilini, L.; Rodrìguez, O.; Pérez, J.M.; Colledani, M. Lithium-ion batteries towards circular economy: A literature review of opportunities and issues of recycling treatments. *J. Environ. Manag.* **2020**, *264*, 110500. [CrossRef] [PubMed]
4. Padhan, A.M.; Hajra, S.; Nayak, S.; Kumar, J.; Sahu, M.; Kim, H.J.; Alagarsamy, P. Triboelectrification based on NiO-Mg magnetic nanocomposite: Synthesis, device fabrication, and energy harvesting performance. *Nano Energy* **2022**, *91*, 106662. [CrossRef]
5. Hajra, S.; Sahu, M.; Sahu, R.; Padhan, A.M.; Alagarsamy, P.; Kim, H.; Lee, H.; Oh, S.; Yamauchi, Y.; Kim, H.J. Significant effect of synthesis methodologies of metal-organic frameworks upon the additively manufactured dual-mode triboelectric nanogenerator towards self-powered applications. *Nano Energy* **2022**, *98*, 107253. [CrossRef]
6. Sidhu, R.K.; Ubhi, J.S.; Aggarwal, A. A Survey Study of Different RF Energy Sources for RF Energy Harvesting. In Proceedings of the International Conference on Automation, Computational and Technology Management, London, UK, 24–26 April 2019; pp. 530–533.
7. Lim, T.B.; Lee, N.M.; Poh, B.K. Feasibility study on ambient RF energy harvesting for wireless sensor network. In Proceedings of the IEEE MTT-S International Microwave Workshop Series on RF and Wireless Technologies for Biomedical and Healthcare Applications, Singapore, 9–11 December 2013; pp. 1–3.
8. Ha, B.W.; Park, J.A.; Jin, H.J.; Cho, C.S. Energy transfer and harvesting for RF-Bio applications. In Proceedings of the IEEE MTT-S International Microwave Workshop Series on RF and Wireless Technologies for Biomedical and Healthcare Applications, Taipei, Taiwan, 21–23 September 2015; pp. 54–55.
9. Sivagami, P.; Pushpavalli, M.; Abirami, P.; Sindhuja, S.; Reddy, N.S. Implementation of RF Energy Harvesting For Mobile Charging. In Proceedings of the IEEE International Conference on Computational Intelligence and Computing Research, Las Vegas, NV, USA, 13–15 December 2018; pp. 1–4.
10. Zhang, H.; Lu, N.; Li, J.; Song, R.; Liu, Y. Lifetime Analysis for Ambient RF Energy Harvesting IoT Node. In Proceedings of the IEEE 5th International Conference on Computer and Communications, Chengdu, China, 6–9 December 2019; pp. 2157–2161.
11. Radhika, N.; Tandon, P.; Prabhakar, T.V.; Vinoy, K.J. RF Energy Harvesting for Self Powered Sensor Platform. In Proceedings of the 16th IEEE International New Circuits and Systems Conference, Montreal, QC, Canada, 24–27 June 2018; pp. 148–151.
12. Ramalingam, L.; Mariappan, S.; Parameswaran, P.; Rajendran, J.; Nitesh, R.M.; Kumar, N.; Nathan, A.; Yarman, B.S. The Advancement of Radio Frequency Energy Harvesters (RFEHs) as a Revolutionary Approach for Solving Energy Crisis in Wireless Communication Devices: A Review. *IEEE Access* **2021**, *9*, 106107–106139. [CrossRef]

13. Khan, D.; Oh, S.; Shehzad, K.; Basim, M.; Verma, D.; Pu, Y.; Lee, M.; Hwang, K.; Yang, Y.; Lee, K. An Efficient Reconfigurable RF-DC Converter With Wide Input Power Range for RF Energy Harvesting. *IEEE Access* **2020**, *8*, 79310–79318. [CrossRef]
14. Chong, G.; Ramiah, H.; Yin, J.; Rajendran, J.; Mak, P.I.; Martins, R.P. A Wide-PCE-Dynamic-Range CMOS Cross-Coupled Differential-Drive Rectifier for Ambient RF Energy Harvesting. *IEEE Trans. Circuits Syst. II Express Briefs* **2019**, *68*, 1743–1747. [CrossRef]
15. Taghadosi, M.; Albasha, L.; Quadir, N.A.; Rahama, Y.A.; Qaddoumi, N. High Efficiency Energy Harvesters in 65nm CMOS Process for Autonomous IoT Sensor Applications. *IEEE Access* **2018**, *6*, 2397–2409. [CrossRef]
16. Eswaran, U.; Ramiah, H.; Kanesan, J.; Rezal, A.W. Design of wideband LTE power amplifier with novel dual stage linearizer for mobile wireless communication. *J. Circuits Syst. Comput.* **2014**, *23*, 1450111. [CrossRef]
17. Shasidharan, P.; Ramiah, H.; Rajendran, J. A 2.2 to 2.9 GHz Complementary Class-C VCOWith PMOS Tail-Current Source Feedback Achieving—120 dBc/Hz Phase Noise at 1 MHz Offset. *IEEE Access* **2019**, *7*, 91325–91336. [CrossRef]
18. Oh, S.; Wentzloff, D.D. −32dBm sensitivity RF power harvester in 130nm CMOS. In Proceedings of the IEEE Radio Frequency Integrated Circuits Symposium, Montreal, QC, Canada, 17–19 June 2012; pp. 483–486.
19. Eswaran, U.; Ramiah, H.; Kanesan, J.; Reza, A. Class-E GaAs HBT power amplifier with passive linearization scheme for mobile wireless communications. *Turk. J. Electr. Eng. Comput. Sci.* **2014**, *22*, 1210–1218. [CrossRef]
20. Hameed, Z.; Moez, K. A 3.2 V–15 dBm Adaptive Threshold-Voltage Compensated RF Energy Harvester in 130 nm CMOS. *IEEE Trans. Circuits Syst. I Regul. Pap.* **2015**, *62*, 948–956. [CrossRef]
21. Chouhan, S.S.; Halonen, K. Threshold voltage compensation scheme for RF-to-DC converter used in RFID applications. *Electron. Lett.* **2015**, *51*, 892–894. [CrossRef]
22. Ouda, M.H.; Khalil, W.; Salama, K.N. Wide-Range Adaptive RF-to-DC Power Converter for UHF RFIDs. *IEEE Microw. Wirel. Compon. Lett.* **2016**, *26*, 634–636. [CrossRef]
23. Ouda, M.H.; Khalil, W.; Salama, K.N. Self-Biased Differential Rectifier with Enhanced Dynamic Range for Wireless Powering. *IEEE Trans. Circuits Syst. II Express Briefs* **2017**, *64*, 515–519. [CrossRef]
24. Shailesh, S.C.; Marko, N.; Kari, H. Efficiency enhanced voltage multiplier circuit for RF energy harvesting. *Microelectron. J.* **2016**, *48*, 95–102.
25. Li, C.; Yu, M.; Lin, H. A Compact 0.9-/2.6-GHz Dual-Band RF Energy Harvester Using SiP Technique. *IEEE Microw. Wirel. Compon. Lett.* **2017**, *27*, 666–668. [CrossRef]
26. Moghaddam, A.K.; Chia, A.; Ramiah, H.; Churchill, K. A self-protected, high-efficiency CMOS rectifier using reverse DC feeding self-body-biasing technique for far-field RF energy harvesters. *AEU—Int. J. Electron. Commun.* **2022**, *152*, 154238. [CrossRef]

 *micromachines*

Article

A 53-µA-Quiescent 400-mA Load Demultiplexer Based CMOS Multi-Voltage Domain Low Dropout Regulator for RF Energy Harvester

Balamahesn Poongan [1], Jagadheswaran Rajendran [1,*], Li Yizhi [1], Selvakumar Mariappan [1], Pharveen Parameswaran [1], Narendra Kumar [2], Masuri Othman [3] and Arokia Nathan [4]

[1] Collaborative Microelectronic Design Excellence Center (CEDEC), Universiti Sains Malaysia, Bayan Lepas 11900, Penang, Malaysia
[2] Department of Electrical Engineering, Faculty of Engineering, University of Malaya, Kuala Lumpur 50603, Wilayah Persekutuan, Malaysia
[3] Institute of Microengineering and Nanoelectronics, Universiti Kebangsaan Malaysia, Bangi 43600, Selangor, Malaysia
[4] Darwin College, Cambridge University, Cambridge CB3 9EU, UK
* Correspondence: jaga.rajendran@usm.my

Abstract: A low-power capacitorless demultiplexer-based multi-voltage domain low-dropout regulator (MVD-LDO) with 180 nm CMOS technology is proposed in this work. The MVD-LDO has a 1.5 V supply voltage headroom and regulates an output from four voltage domains ranging from 0.8 V to 1.4 V, with a high current efficiency of 99.98% with quiescent current of 53 µA with the aid of an integrated low-power demultiplexer controller which consumes only 68.85 pW. The fabricated chip has an area of 0.149 mm^2 and can deliver up to 400 mA of current. The MVD-LDO's line and load regulations are 1.85 mV/V and 0.0003 mV/mA for the low-output voltage domain and 3.53 mV/V and 0.079 mV/mA for the high-output voltage domain. The LDO consumes only 174.5 µW in standby mode, making it suitable for integrating with an RF energy harvester chip to power sensor nodes.

Keywords: multi-voltage domain low dropout; radio frequency energy harvester; power management unit; bandgap voltage reference; system on chip; internet of things; feedback network; demultiplexer

Citation: Poongan, B.; Rajendran, J.; Yizhi, L.; Mariappan, S.; Parameswaran, P.; Kumar, N.; Othman, M.; Nathan, A. A 53-µA-Quiescent 400-mA Load Demultiplexer Based CMOS Multi-Voltage Domain Low Dropout Regulator for RF Energy Harvester. *Micromachines* **2023**, *14*, 379. https://doi.org/10.3390/mi14020379

Academic Editors: Qiongfeng Shi and Huicong Liu

Received: 27 November 2022
Revised: 14 January 2023
Accepted: 15 January 2023
Published: 2 February 2023

1. Introduction

A radio frequency energy harvester (RFEH)–based system on chip (SoC) is an integrated circuit that is capable of operating with much less dependency on battery energy [1]. It is made up of many different sub-circuit blocks, such as analog and mixed signal blocks, and each one requires a different voltage domain from the power source, which has high power quality requirements. Hence, RFEH can extend the life of battery-powered SoCs.

A wireless sensor node (WSN), on the other hand, which is made up of four major components—a sensing unit, a power unit, a processing unit, and a transmitter—requires a better power solution to extend the battery life of the sensor, which continuously monitors, captures, and transmits data to the processing unit for processing [2,3]. Each wireless sensor node component operates in a different voltage domain.

Besides this, a series of prominent microcontrollers widely used in the industry still require a dynamic voltage scaling power solution. The microcontrollers have a supply voltage range that varies based on the system frequency. For example, flash memory programming is not necessary for microcontrollers, such as MSP430G2001, when the system frequency is 1 MHz. During this phase, the supply voltage range is from 1.8 V to 3.6 V. Meanwhile, the supply voltage range is from 2.2 V to 3.6 V if flash memory programming is needed [2,4]. The static or single-power solution is not helpful in such conditions. A notable power solution is required for the system to increase its power efficiency.

A single-inductor multiple-output (SIMO) device that switches out multiple independent DC/DC converters could be the solution for multi-voltage domains [5–7]. The SIMO DC/DC converter can regulate multiple outputs while maintaining high power efficiency thanks to a single internal inductor component. The SIMO approach has an advantage over traditional DC/DC regulators in that it consumes less power and occupies less chip area than traditional methods, which necessitate the use of multiple independent regulators to supply multiple domains. Although this strategy helps to increase power efficiency, large off-chip inductors prevent its use in SoC architecture because they take up a lot of space. Aside from that, output ripples, switching noise, and cross regulation from the switching regulator, as well as the failure to provide a clean supply to the device, are significant flaws in this solution.

In the last decade, the power management integrated circuit (PMIC), consisting of a switching regulator and several low-dropout regulators, has sparked much interest as a multi-voltage domain solution. Even though switching regulators do not provide clean power, they are more efficient than LDOs in terms of the massive voltage drop from input to output. Because the switching regulator's output must be ripple-free, the LDO is a good solution for providing clean power because of its high power supply rejection capability. Both regulators were built as a single integrated circuit to improve power efficiency and reduce ripple in the regulated voltage. LDOs are placed in the back end and switching regulators are placed in the front end. However, when used for an embedded or SoC solution, the complexity of fabricating inductors for PMIC is a significant disadvantage [8]. Furthermore, PMIC is better suited for high-power systems than low-power solutions.

Because LDO can provide a clean supply with a strong supply rejection ability, it is a popular alternative to PMIC for supplying different voltages for SoC system applications [9]. However, a large external capacitor was placed at the output LDO to reduce overshoot and undershoot while maintaining stability. This is difficult for SoC applications due to space constraints. The use of capacitorless, OCL-LDO to overcome area constraints is ideal for portable electronic devices and SoC power delivery applications. However, to support multiple voltage domain operations in SoC, typically multiple LDOs are needed to accommodate in the SoC, which consumes more area on the SoC chip. Furthermore, power efficiency in numerous independent LDO have deteriorated because all LDO consume power continuously, even if some voltage domains only operate infrequently and do not require constant supply [10].

In addition to multiple independent LDO, researchers are interested in designing LDO with programmable output voltage [11–13]. The output voltage was varied by changing the ratio of the feedback resistor via the transistor switch. The control signal activated and deactivated the transistor switch. The programmable output voltage topologies in the cited paper [11–13] used control logic, consisting of many CMOS switches in the feedback network, to vary the output voltage. Various topologies such as multi-voltage control, adaptive reference control, dynamic voltage, and frequency scaling were used to control the gate voltage of the CMOS switches to achieve the desired output voltage of LDO. These circuits consume a large area on the chip.

J. H. Wang et al. proposed a programmable 32-step output voltage that modifies the digital pulse width modulation (DPWM) clock and supplies it with a clean supply voltage [14]. The output voltage was modified using digital LDO topology in this technique. The phase frequency detector detected the clock differences between the reference clock and DPWM output clock. The output clock of DPWM was fed into the phase-frequency detector, PFD. When the output clock of DPWM differs from the reference clock in frequency or phase, the PFD captures it and generates a voltage signal to indicate the variation in phase or frequency. The main objective of the TDC employed in this design is to convert the output of the PFD signal into a 5-bit digital signal, which is then used to correct the output clock of the DPWM by adjusting the output voltage of the proposed LDO regulator, as explained in reference [14].

A higher resolution of step size is also necessary for DLDO to ensure that the output voltage is regulated with high accuracy. Moderate step size or step resolution leads to drawbacks in output voltage accuracy [14]. However, high power consumption and a large chip area are the trade-offs for the high-resolution step circuit. Hence, a technique for programmable reference voltage with a wider input common mode error amplifier has been introduced in addition to programmable output voltage LDO [15,16].

To address the issues above, the MVD-LDO has been proposed in this work. The proposed MVD-LDO has an on-chip capacitor and can handle up to 400 mA load current, making it suitable for embedded and SoC applications. Sections 2 and 3 describe the application and circuit architecture of the proposed multi-voltage domain low dropout regulator with demultiplexer, followed by Section 4, which elaborates on circuit implementation of multi-voltage domain LDO. Section 5 displays the measurement results of the proposed design, and Section 6 presents the conclusion.

2. MVD-LDO for RFEH Based IoT Systems

The proposed MVD- LDO is suitable for RFEH-based IoT System on Chip (SoC). The SoC's basic building blocks are processor units, sensors, transmitters, and receivers. Because the SoC contains numerous devices, different voltage rails were required to power the SoC. Even if each device serves a distinct purpose, it is not necessary to keep them all turned on at all times because this reduces battery life. As a result, MVD-LDO is a viable alternative to conventional multi-LDO, which consumes more power, costs more, and takes up more space. Figure 1 depicts a block diagram of an MVD-LDO for RFEH-based Internet of Things applications.

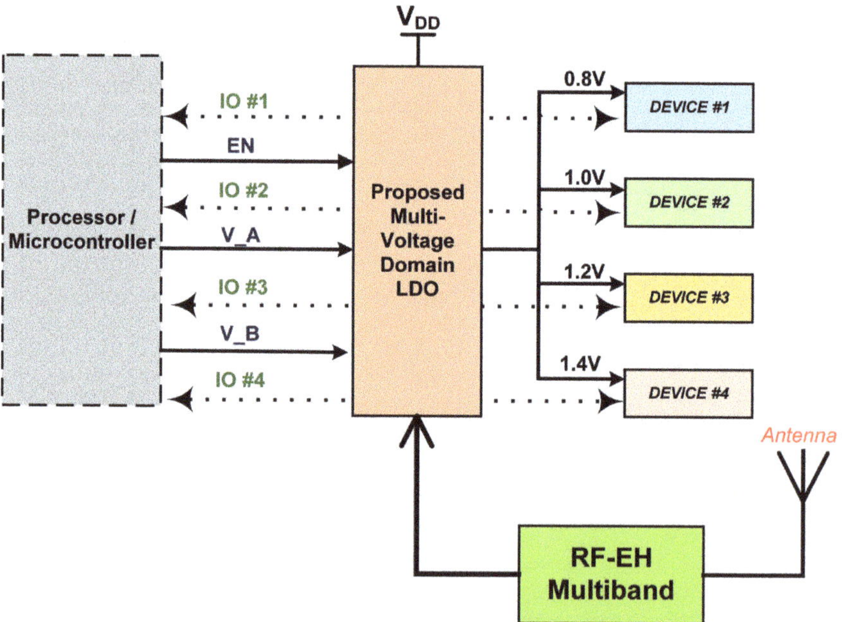

Figure 1. Block diagram of MVD-LDO for RFEH-based IoT applications.

To supply the source, the MVD-LDO's output is distributed to all the devices, and each device's IO pin is connected to the processing unit to signal the device's state while it is powered up. The processing unit will initially configure the LDO to output the voltage following the sequence. Once the work is completed, the first device sends an acknowledgment signal to the processor via an IO pin. In response to the acknowledgement from the previous device, the processor determines which device should be turned on next.

As a result, the MVD-LDO is used to implement a power-efficient supply mechanism on the SoC.

3. Circuit Architecture of MVD-LDO

Figure 2 illustrates the schematic of the MVD-LDO which is capable of producing up to four regulated output voltages (Vout 1, 2, 3, 4) with a DEMUX controller.

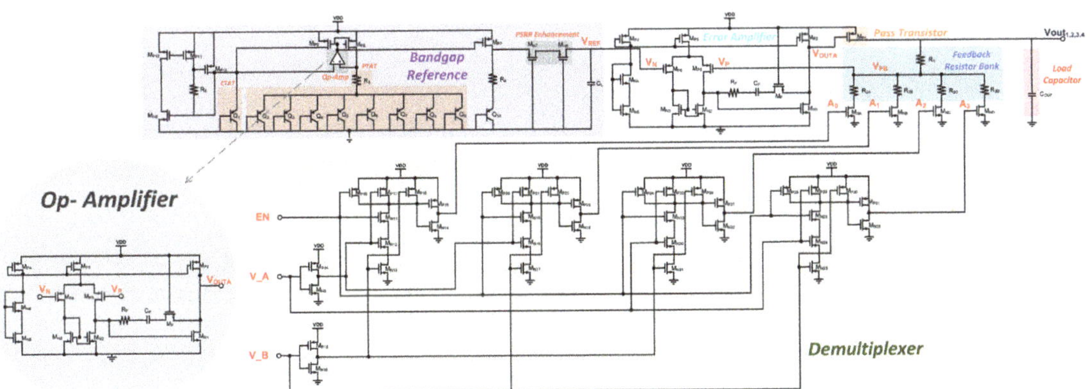

Figure 2. The circuit architecture of the MVD-LDO.

This MVD- LDO regulator comprises a set of feedback resistors, a pass element, a low-power error amplifier, bandgap reference voltage (BGR), demultiplexer, and on-chip load capacitor. A resistive divider of the output voltage generates the feedback voltage, V_{FB}, which varies proportionally with the output voltage. A reference voltage, V_{REF}, is supplied through the bandgap reference (BGR) circuit. When the feedback voltage, V_{FB}, deviates from the reference voltage, V_{REF}, the error amplifier is used to correct the error at the output voltage by adjusting the gate voltage of the pass device. The error amplifier corrects errors by varying the pass transistor's gate voltage and assuring the regulated voltage is within design parameters.

The feedback circuit in the proposed MVD-LDO consists of R_1, R_{2A}, R_{2B}, R_{2C}, and R_{2D} to provide four output voltages. In addition, for switching purposes, an NMOS transistor has been added to each pathway of the feedback resistor. In response to user input, the demultiplexer activated the gate voltage of an NMOS transistor. Only one output voltage can be regulated at any given time. The constituent units are discussed in the following sections.

4. Circuit Implementation of MVD-LDO

Figure 3 depicts a single voltage domain LDO with the output voltage controlled by feedback resistors or the reference voltage of the op-amp [17].

The LDO's V_{out} is given as:

$$V_{out} = \left(1 + \frac{R_1}{R_2}\right) V_{FB} \tag{1}$$

where V_{FB}, is the feedback voltage of the single voltage domain LDO. The error amplifier regulates the gate voltage of the pass transistor by comparing V_{FB} to the reference voltage, V_{REF}. If an error amplifier's input is unequal, the pass transistor's gate voltage is varied to control the required output voltage. This activity loops indefinitely to ensure that the output voltage is accurately regulated.

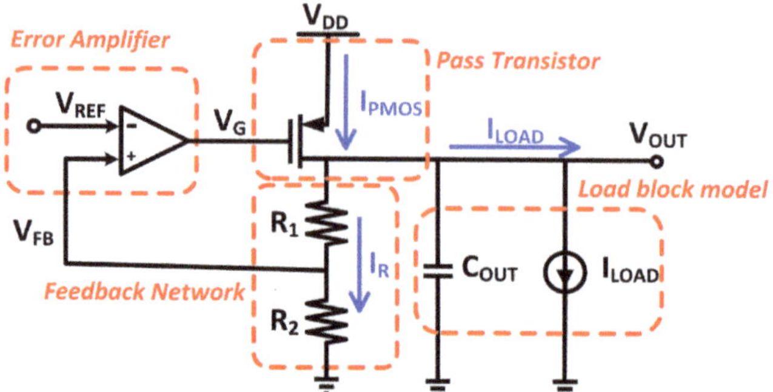

Figure 3. Conventional LDO regulator.

Off-chip placement of the feedback resistor is common, and the combination of feedback resistor values is positioned according to the desired regulated output voltage. Unfortunately, using an off-chip approach raises the design cost and is unsuitable for embedded design. Aside from that, an incorrect resistor value influences the regulator's current efficiency, stability, and other critical parameters, resulting in a system with poor power efficiency.

Altering the reference voltage, V_{REF}, while preserving the feedback resistor's ratio is an additional way for altering the output voltage [17]. Since the reference voltage usually supplied from the bandgap reference, which is conspicuous across the process, voltage, and temperature, and tweaking the reference voltage is not a good solution to achieve acceptable output voltage. Hence, in MVD-LDO, adjusting the feedback resistor ratio has been adopted to govern the multi-voltage domain. Figure 4 displays the design of the MVD-feedback LDO's resistor.

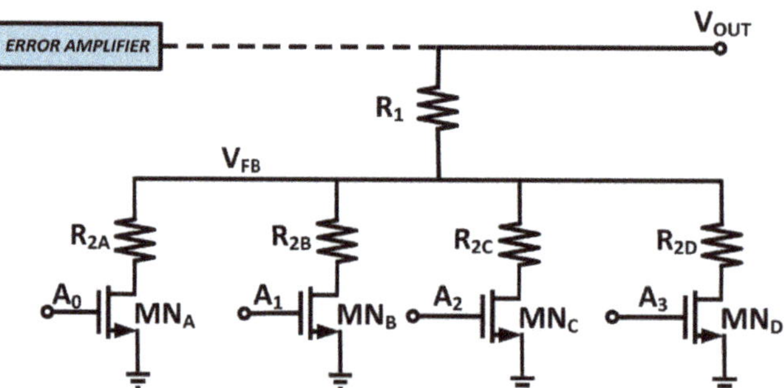

Figure 4. Feedback resistor design of MVD-LDO.

In our work, the bottom resistor of the feedback network circuit has been developed using a resistor bank, which increases the precision of the output voltage and offers multiple V_{out}. Since this design intends to handle four V_{out}, a total of four resistors, R_{2A}, R_{2B}, R_{2C}, and R_{2D}, have been added to the bottom of the feedback resistor. NMOS transistors MN_A, MN_B, MN_C, and MN_D serve as switches. At a given moment, only one path is activated. In accordance with the ratio of the feedback resistor in the active circuit, the output voltage is regulated. An on-chip low power consumption demultiplexer design has been implemented to manage the switches and adjust the output voltage correspondingly. The power consumption of the DEMUX is only 68.85 pW. This is considered in the worst-case scenario, when all three inputs of the DEMUX have been turned on.

The digital 2:4 demultiplexer has been utilized on the proposed MVD-LDO. This demultiplexer uses three input signals, which are one enable pin and two logic input pins, to generate four output signals that regulate the feedback network's switched-on state. A_0,

A_1, A_2, and A_3 are the demultiplexer's output signals, whereas V_A, V_B, and EN are the input signals

Similar to V_{DD}, the input voltage of the demultiplexer is 1.5 V. The enable pin, EN, is employed to activate the LDO regulator. If the EN signal logic is LOW, the DEMUX is powered off as all four switches are turned off. Therefore, V_{FB} is the same as V_{DD}. As a result, the gate voltage of the pass devices rises as the error amplifier's output voltage increases, resulting in the eventual shutdown of the LDO. The shutdown process puts the SoC into sleep mode, critical for conserving battery energy.

To achieve low power consumption, the W/L of the demultiplexer's transistors are optimized with reference to (2):

$$I_{D,sat} = \frac{\mu_p . C_{ox}}{2} . \frac{W}{L} . (V_{GS} - V_T)^2 \tag{2}$$

As a result, the demultiplexer consumes only 30.5 pA and 45.9 pA of current in standby and operating modes, respectively. The EN is used to toggle between operating and standby mode. The simulation results of the current consumption of the demultiplexer during standby and operating modes are shown in Figure 5. The simulation was performed on the Cadence Virtuoso platform with Silterra CMOS 180 nm process technology.

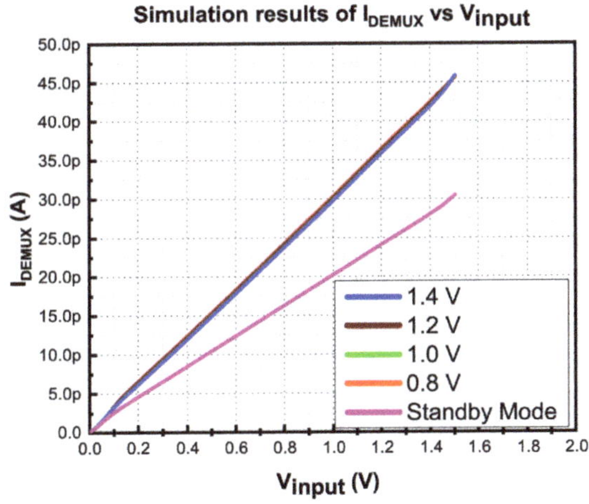

Figure 5. Simulation results of MVD-DEMUX's current consumption.

Most of the logic gate was shut down during standby mode since the EN signal is low. Therefore, it consumes minimal current. On the other hand, when all logic is enabled, the current consumption rises rapidly. In the proposed MVD-LDO, the EN, V_A, and V_B must be set to regulate the 1.4 V. This circumstance causes the DEMUX to consume more current.

Additionally, the MVD-LDO was also designed with high current efficiency. The current efficiency of the MVD-LDO is dependent on the load and quiescent current. Its current efficiency is expressed as (3):

$$Current Efficiency, \% = \frac{I_{load}}{I_{load} + I_{bgr} + I_{EA} + I_{DEMUX} + I_R} \times 100\% \tag{3}$$

where I_{load} is the load current, I_{bgr} is the BGR current, I_{EA} is the error amplifier current, I_{DEMUX} is the demultiplexer current, and I_R is the current flowing into the feedback resistor bank. Optimizing the W/L of the transistors in the BGR, error amplifier, and the demultiplexer reduces the quiescent current without degrading other critical performance factors. I_R is reduced with reference to (4):

$$I_R = \frac{V_{REF}}{R} \tag{4}$$

where the V_{REF} is the reference voltage, and reduction of the quiescent current increases the current efficiency [18].

The simulated quiescent current of the MVD-LDO is plotted in Figure 6.

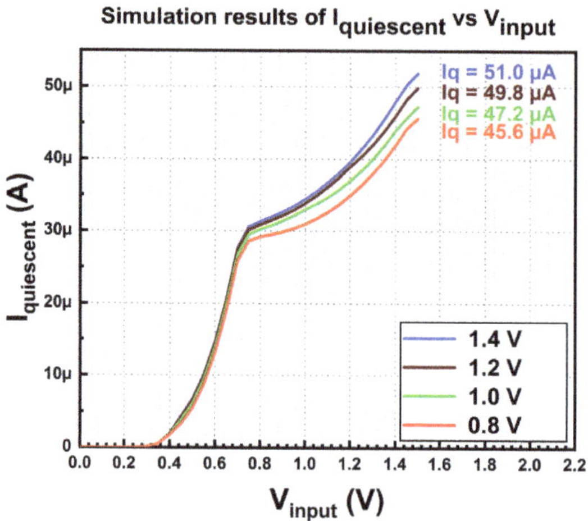

Figure 6. Simulation results of the MVD-LDO's quiescent current.

The MVD-LDO is designed to carry 400 mA of load current. To accomplish this, parallel PMOS transistors operating in the saturation region were implemented as pass transistors [19], as depicted in Figure 7. The size of the pass transistor is a matter to support the large current. The load current of the LDO is usually determined by the capacity of the pass device.

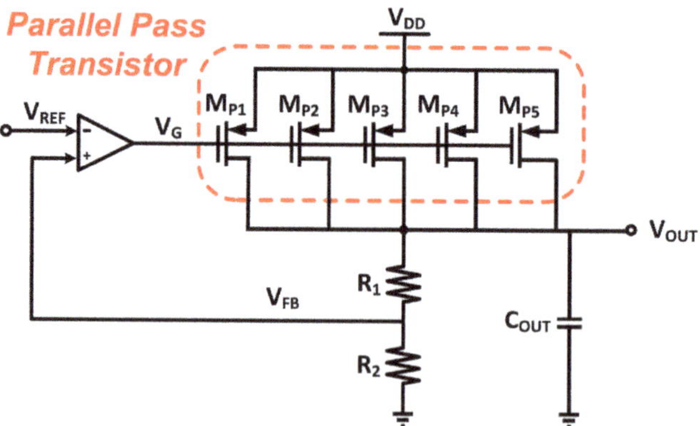

Figure 7. Parallel pass transistor design.

Smaller pass devices can only support a modest load current, whereas larger devices can support a large current based on the transistor's capacity. By increasing the size of the transistor to support higher load current, it impacts the size of the chip. In the proposed MVD-LDO, multiple transistors are connected in parallel, while the size of each transistor is significantly smaller. The parallel pass transistor approach increases the capacity of the pass device to accommodate large load currents while keeping the transistor size smaller. The parallel PMOS also significantly reduces the quiescent current. Consequently, LDO's current efficiency is enhanced.

The load regulation of the MVD-LDO is related to closed-loop DC output resistances, $R_{out,cl}$, of the LDO, as illustrated in (5):

$$LR = \frac{\Delta V_{out}}{\Delta I_{out}} = R_{out,cl} \cong \frac{R_{out}}{1 + \beta g_{mp} R_{out} A_{EA,O}} \tag{5}$$

where g_{mp} is the transconductance of the pass transistor, R_{out} is the output impedance, β is the feedback factor of the amplifier, and $A_{EA,O}$ is the DC gain of the error amplifier.

To achieve good load regulation, the closed-loop DC output resistance, $R_{out,cl}$ was designed to be smaller [20]. Referring to (9), the $R_{out,cl}$, reduces when the DC gain of the error amplifier, $A_{EA,O}$, increases. Figure 8 illustrates the high gain op-amp used as error amplifier in the MVD-LDO.

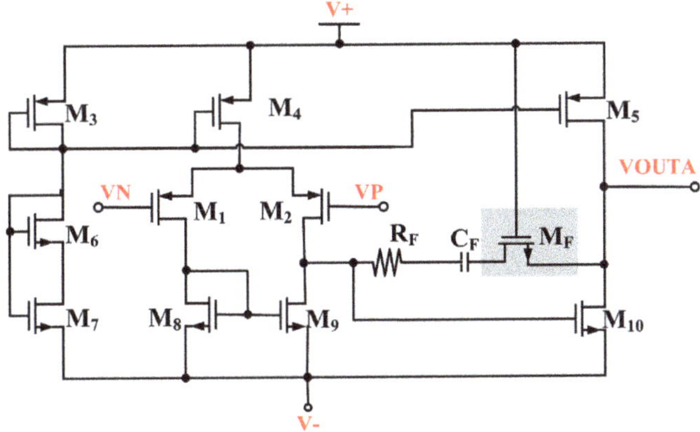

Figure 8. Schematic of the high gain error amplifier.

The high gain error amplifier consists of an operational transconductance amplifier as first stage and a common source amplifier as second stage. The common source delivers higher output voltage swing. M_1 and M_2 are the input stage which operates in the saturation region. The equivalent transfer function is given as:

$$H(s) = \frac{K(1 + \frac{s}{\omega_z})}{(1 + \frac{s}{\omega_{p1}})(1 + \frac{s}{\omega_{p2}})} \tag{6}$$

The overall DC gain of the error amplifier is given as:

$$DCGain = gm_{M1} * gm_{M10} * (ro_{M2}//ro_{M9}) * (ro_{M5}//ro_{M10}) \tag{7}$$

The MVD-LDO achieves optimum load regulation and transient analysis by boosting the error amplifier's DC gain. The transient analysis of the LDO is simulated and plotted in Figure 9.

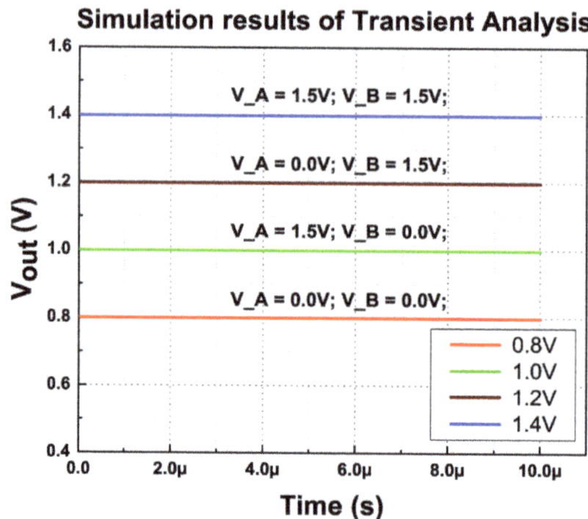

Figure 9. Simulation results of the MVD-LDO's transient analysis.

Based on the transient analysis simulation, the output voltage ripple was obtained. It is evident that without the external capacitor, the output voltage variation is significantly lower. Figure 10, employed the simulation results of the output voltage ripple for all four outputs during full-load conditions.

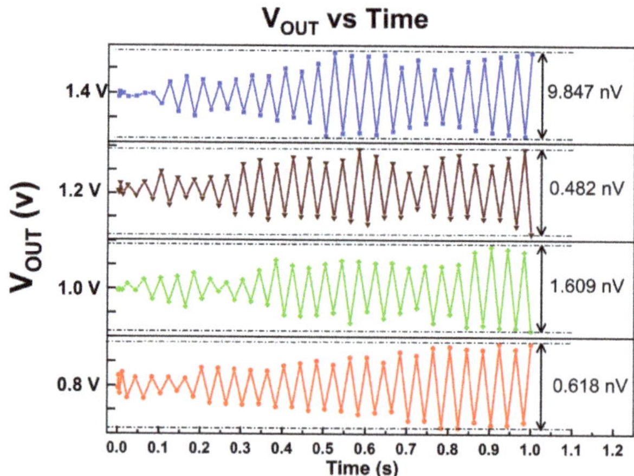

Figure 10. Simulated output voltage ripple of proposed MVD-LDO.

5. Measurement Results

The proposed multi-voltage domain LDO regulator was fabricated using 180 nm CMOS process technology to verify the feasibility of the proposed design topology. Figure 11 depicts the micrograph of the proposed LDO.

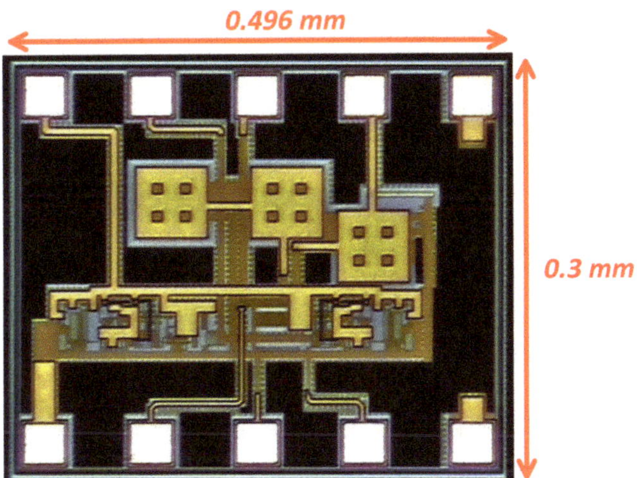

Figure 11. Micrograph of the proposed MVD-LDO.

The proposed LDO regulator can support a load current of up to 400 mA without using any external component, especially an off-chip capacitor (C_L). Figure 12 illustrates the measured multi-voltage domain output voltage from 0.8 V to 1.4 V, with different configuration during no load conditions. The test configuration setup with digital logic is presented in Table 1.

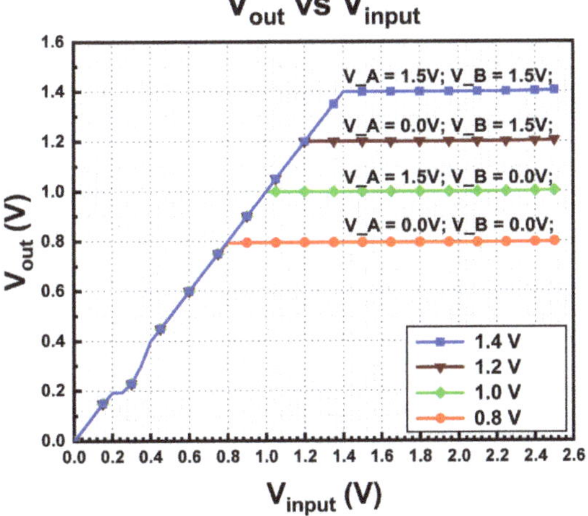

Figure 12. Output voltage of proposed MVD-LDO.

The test configuration setup with digital logic is presented in Table 1.

Table 1. Test configuration of MVD-LDO regulator.

V_{out}, V	V_A	V_B
0.8	Low	Low
1.0	Low	High
1.2	High	Low
1.4	High	High

Line regulation of LDO computes from the measured result from Figure 12 when the supply voltage is 1.5 V. For output voltage of 0.8 V, the line regulation is 1.85 mV/V, while for 1.0 V regulated voltage, the line regulation is 2.7 mV/V. Besides this, the line regulation values for output voltages of 1.2 V and 1.4 V are 2.19 mV/V and 3.53 mV/V, respectively.

The transient analysis of the measured waveform is shown in Figure 13. All four regulated output voltages look stable.

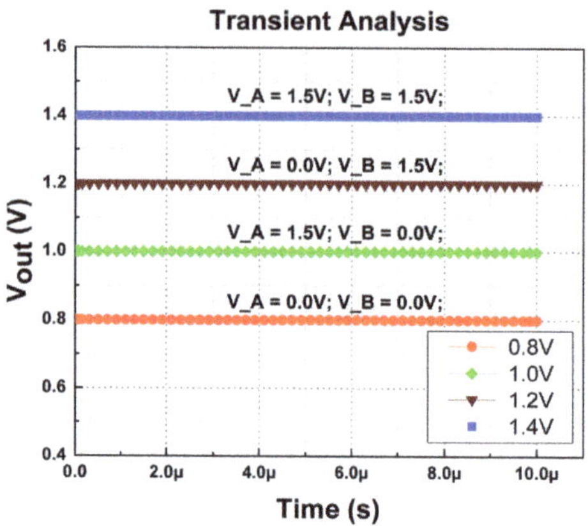

Figure 13. Transient analysis of proposed MVD-LDO.

Figure 14 demonstrates the quiescent current of the multi-voltage domain LDO regulator for each test configuration. During this measurement, the regulator connects to zero load. Based on the measured results, the lowest quiescent current is 46.8 µA, while the highest is 52.9 µA with the output voltage of 0.8 V and 1.4 V, respectively. The highest multi-voltage domain output voltage contributes to the largest quiescent current, while otherwise, the relationship is reversed. The results dictate that output voltage is directly proportional to quiescent current. This quiescent current is the total operating current for bandgap reference, differential amplifier, and demultiplexer, which operate during no-load and full-load conditions.

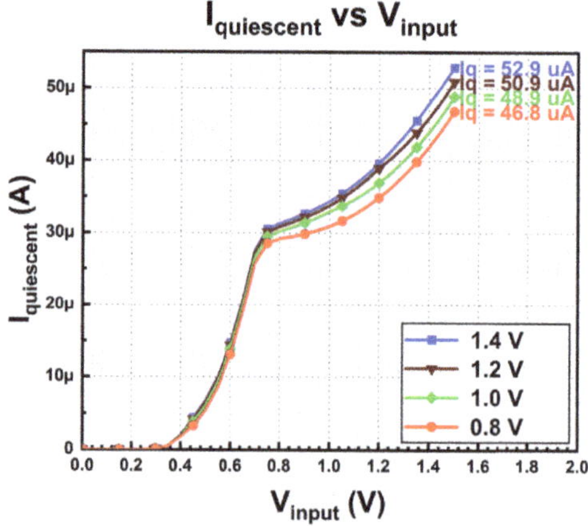

Figure 14. Quiescent current of proposed MVD-LDO.

The load regulation on the multi-voltage domains is shown in Figure 15. The output voltage was measured while varying the load from no-load condition to full-load condition with current step of 20 mA. The load regulation values for multi-voltages of 0.8 V, 1.0 V, 1.2 V, and 1.4 V are 0.0003 mV/mA, 0.000325 mV/mA, 0.0004 mV/mA, and 0.079 mV/mA, respectively. Like line regulation, the load regulation has the smallest load regulation when the lowest multi-voltage domain voltage is highest when the output voltage is huge.

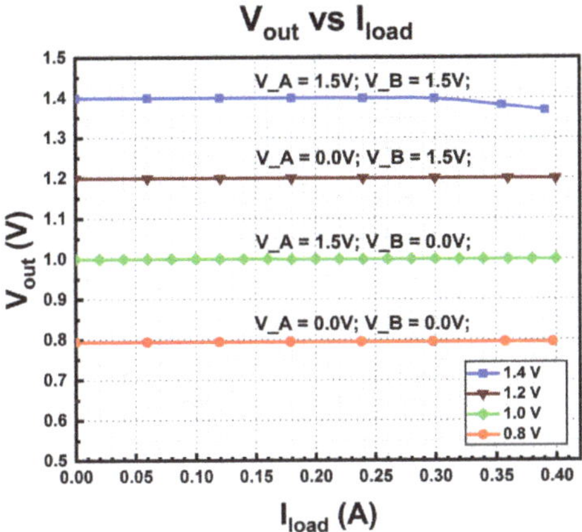

Figure 15. Measured load regulation of proposed MVD-LDO.

Current efficiency is an important parameter in LDO regulators as the input and output current are close enough. To evaluate the current efficiency of the regulator, the input current and output current were measured during the full-load condition. Based on the measured results, the current efficiency of the regulator for all four output voltages is computed and displayed in Table 2.

Table 2. Current efficiency of proposed MVD-LDO.

V_{output}, V	I_{output}, mA	I_{input}, mA	Power Efficiency, %
0.8	399.5	399.6	99.98
1.0	399.5	399.8	99.98
1.2	399.7	399.8	99.98
1.4	397.0	397.1	99.98

The current efficiency of the MVD-LDO regulator was about 99.98% for all four tested configurations.

Table 3 summarizes the performance of the proposed multi-voltage domain LDO regulator and compares it with other state-of-the-art LDOs.

Table 3. Summarized performance of proposed MVD-LDO compared with other state-of-the-art LDOs.

Reference	[8]	[21]	[22]	[23]	[24]	This Work
Year	2021	2017	2020	2020	2019	2022
Topology	Multi Domain LDO	LDO	LDO	LDO	DLDO	Multi Domain LDO
Technology (nm)	180	130	40	180	65	180
Chip Area (mm^2)	0.49	0.1825	None	0.037	None	0.149
Input Voltage, V_{IN}(V)	3.3–3.6	1.05–2.0	1.1–1.9	1.4–1.8	0.8–1.2	1.5
Output Voltage, V_{IN}(V)	3.2 (2.4, 1.6,0.8) [1]	1.0	0.2–1.1	1.2–1.6	0.6–1.15	(0.8;1.0;1.2;1.4) [1]
Dropout voltage, V_{DO}(mV)	100	29.7	200	200	200	100–700
Load Current, I_{load}(mA)	50	300	100	300	50	400
Quiescent Current, I_q(µA)	239	14–120	56	0.94–255	26.25–105	46.83–52.88
Current efficiency, %	96.5	99.96	99.94	99.92	None	99.98
Load regulation (mV/mA)	(V_{OUTH}: 4.38×10^{-7}; V_{OUTL}: 3.13×10^{-6}) [2];	0.006	0.176 @ $V_{OUT} = 0.2$; 0.2 @ $V_{OUT} = 1.1$	0.1	None	0.0003–0.079;
Line regulation (mV/V)	1.13	0.44	0.857@ $V_{OUT} = 0.2$; 5.0 @ $V_{OUT} = 1.1$	5.33	None	1.85–3.53
Output Capacitor, C_O (µF)	None	1	1	1	0.1	None

[1] Multi-voltage domain. [2] Unit for load regulation is %/mA.

6. Conclusions

This paper proposes a capacitorless multi-voltage domain low-dropout regulator (MVD-LDO) in 180 nm CMOS. The MVD-LDO can operate at 1.5 V, encompass a broad output range of 0.8V to 1.2V, and provide 400 mA to each of its four output voltage domains. A low-power integrated demultiplexer realizes this efficient multiple output voltage. In addition, the MVD-LDO draws 46.83 µA for lower output and 52.88 µA for higher output in the absence of load. The design of the error amplifier was optimized with a high DC gain, resulting in outstanding load and line regulation of 0.001 mV/mA and 1.85 mV/V, respectively. With a load of 400 mA and across the entire voltage domain, the current efficiency was 99.98%, making this an efficient power management component unit for RF energy harvesters.

Author Contributions: Conceptualization, B.P. and J.R.; methodology, B.P, J.R., L.Y., S.M., and P.P; software, B.P., L.Y., and P.P.; formal analysis, B.P., J.R., S.M., and P.P.; validation, B.P., J.R., S.M., N.K., and M.O.; resources, J.R. and S.M.; data curation, B.P., J.R., and P.P.; writing—original draft preparation, B.P., J.R., L.Y., and S.M.; writing— review and editing, B.P., J.R., S.M., N.K., M.O., and A.N.; visualization, B.P., L.Y., and P.P.; supervision, J.R., S.M., and M.O.; project administration, J.R.; funding acquisition, J.R. and S.M. All authors have read and agreed to the published version of the manuscript.

Funding: Malaysian Ministry of Higher Education's Fundamental Research Grant Scheme [grant number: FRGS/1/2019/TK04/USM/02/14] and CREST Malaysia [grant PCEDEC/6050415].

Institutional Review Board Statement: Not applicable.

Informed Consent Statement: Not applicable.

Data Availability Statement: Not applicable.

Acknowledgments: The authors are grateful to Collaborative Microelectronic Design Excellence Center (CEDEC), USM, and Silterra Malaysia, for the laboratory and chip fabrication assistance respectively. Besides that, would like to acknowledge Malaysian Ministry of Higher Education and CREST Malaysia for support us by funding this research.

Conflicts of Interest: The authors declare no conflict of interest.

Abbreviations

The following abbreviations are used in this manuscript:

LDO	Low dropout
MVD-LDO	Multi-voltage domain LDO
RFEH	Radio frequency energy harvester
DLDO	Digital low dropout
BGR	Bandgap reference
EA	Error amplifier

References

1. Ramalingam, L.; Mariappan, S.; Parameswaran, P.; Rajendran, J.; Nitesh, R.M.; Kumar, N.; Nathan, A.; Yarman, B.S. The Advancement of Radio Frequency Energy Harvesters (RFEHs) as a Revolutionary Approach for Solving Energy Crisis in Wireless Communication Devices: A Review. *IEEE Access* **2021**, *9*, 106107–106139. [CrossRef]
2. Zhao, A.; Wang, L.; Yao, C.H. Research on electronic-nose application based on wireless sensor networks. *Chin. J. Electron.* **2006**, *48*, 250–254. [CrossRef]
3. Gao, Y.; Sun, G.; Li, W.; Pan, Y. Wireless sensor node design based on solar energy supply. In Proceedings of the 2009 2nd International Conference on Power Electronics and Intelligent Transportation System (PEITS), Shenzhen, China, 19–20 December 2009; Volume 3, pp. 203–207.
4. Hudgins, D. Dynamic Voltage Scaling with a Dual LDO. 2014, pp. 1–22. Available online: www.ti.com (accessed on 5 January 2023).
5. Ma, Y.S. A low quiescent current and cross regulation single-inductor dual-output converter with stacking MOSFET driving technique. In Proceedings of the ESSCIRC 2017–43rd IEEE Eur. Solid State Circuits Conference, Leuven, Belgium, 11–14 September 2017; Volume 3, pp. 352–355.
6. Zheng, Y.; Guo, J.; Leung, K. A Single-Inductor Multiple-Output Buck-Boost DC–DC Converter with Duty-Cycle and Control-Current Predictor. *IEEE Trans. Power Electron.* **2020**, *35*, 12022–12039. [CrossRef]
7. Qu, Y.; Wang, Z. Soft-Switching Techniques for Single-Inductor. *IEEE Trans. Power Electron.* **2020**, *35*, 13748–13756. [CrossRef]
8. Guo, Z.; Li, D.; Zhang, B.; Xue, Z.; Dong, L.; Chen, Z.; Xiong, Y.; Geng, L. Highly Efficient Fully Integrated Multivoltage-Domain Power Management with Enhanced PSR and Low Cross-Regulation. *IEEE Trans. Power Electron.* **2021**, *36*, 11469–11482. [CrossRef]
9. Kim, Y.; Lee, S. Fast transient capacitor-less LDO regulator using low-power output voltage detector. *Electron. Lett.* **2012**, *48*, 175–177. [CrossRef]
10. Amin, S.; Mercier, P. MISIMO: A multi-input single-inductor multi-output energy harvesting platform in 28-nm fdsoi for powering net-zero-energy systems. *IEEE J. Solid-State Circuits* **2018**, *53*, 3407–3419. [CrossRef]
11. Wu, Y.; Huang, C.; Liu, B. A low dropout voltage regulator with programmable output. In Proceedings of the 4th IEEE Conference on Industrial Electronics and Applications, ICIEA 2009, Xian, China, 25–27 May 2009; pp. 3357–3361.
12. Shen, J.; Yang, W.; Hsieh, C.; Lo, Y. A low power multi-voltage control technique with fast-settling mechanism for low dropout regulator. In Proceedings of the ISIC-2009–12th International Symposium on Integrated Circuits, Singapore, 14–16 December 2009; pp. 558–561.
13. Guochen, A.; Zhanyou, S. Programmable Voltage Regulator Design based on Digitally Controlled Potentiometer. In Proceedings of the 2007 8th International Conference on Electronic Measurement and Instruments, Xian, China, 16–18 August 2007; pp. 1–456.
14. Wang, J.H.; Tsai, C.H.; Lai, S.W. A low-dropout regulator with tail current control for DPWM clock correction. *IEEE Trans. Circuits Syst. II Express Briefs* **2012**, *59*, 45–49. [CrossRef]
15. Tseng, C.Y.; Wang, L.W.; Huang, P.C. An integrated linear regulator with fast output voltage transition for dual-supply SRAMs in DVFS systems. *IEEE J. Solid-State Circuits* **2010**, *45*, 2239–2249. [CrossRef]
16. Zheng, C.; Ma, D. Design of monolithic CMOS LDO regulator with D2 coupling and adaptive transmission control for adaptive wireless powered bio-implants. *IEEE Trans. Circuits Syst. I Regul. Pap.* **2011**, *58*, 2377–2387. [CrossRef]
17. Mo, H.; Kim, D. Multiple-output LDO regulator applying with constant feedback factor. In Proceedings of the 2017 International SoC Design Conference (ISOCC), Seoul, Republic of Korea, 5–8 November 2017; pp. 194–195.
18. Rincon-mora, G.A.; Allen, P.E. A Low-Voltage, Low Quiescent Current, Low Drop-Out Regulator. *IEEE J. Solid-State Circuits* **1998**, *33*, 36–44. [CrossRef]
19. Akram, M.A.; Hwang, I.C.; Ha, S. Architectural Advancement of Digital Low-Dropout Regulators. *IEEE Access* **2020**, *8*, 137838–137855. [CrossRef]
20. Torres, J. Low drop-out voltage regulators: Capacitor-less architecture comparison. *IEEE Circuits Syst. Mag* **2014**, *14*, 6–26. [CrossRef]
21. Duong, Q.; Nguyen, H.; Kong, J. Multiple-loop design technique for high-performance low dropout regulator. *IEEE J. Solid-State Circuits* **2017**, *52*, 2533–2549. [CrossRef]
22. Huang, C.H.; Liao, W.C. A High-Performance LDO Regulator Enabling Low-Power SoC with Voltage Scaling Approaches. *IEEE Trans. Very Large Scale Integr. Syst.* **2020**, *5*, 1141–1149. [CrossRef]

23.	Jeon, I.; Guo, T.; Roh, J. 300 mA LDO Using 0.94-µA IQ with an Additional Feedback Path for Buffer Turn-off under Light-Load Conditions. *IEEE Access* **2021**, *9*, 51784–51792. [CrossRef]
24.	Lu, Y.; Chen, F.; Mok, P.K.T. A Single-Controller-Four-Output Analog-Assisted Digital LDO with Adaptive-Time-Multiplexing Control in 65-nm CMOS. In Proceedings of the ESSCIRC 2019–IEEE 45th Eur. Solid State Circuits Conference, Cracow, Poland, 23–26 September 2019; pp. 289–292.

 *micromachines*

Article

Self-Sustainable IoT-Based Remote Sensing Powered by Energy Harvesting Using Stacked Piezoelectric Transducer and Thermoelectric Generator

Wasim Dipon *, Bryan Gamboa, Maximilian Estrada, William Paul Flynn, Ruyan Guo and Amar Bhalla

Department of Electrical and Computer Engineering, University of Texas at San Antonio, San Antonio, TX 78249, USA; zzn341@my.utsa.edu (B.G.); maximilian.estrada@my.utsa.edu (M.E.); paul.flynn@my.utsa.edu (W.P.F.); ruyan.guo@utsa.edu (R.G.); amar.bhalla@utsa.edu (A.B.)
* Correspondence: wasim.dipon@my.utsa.edu

Abstract: We propose a self-powered remote multi-sensing system for traffic sensing which is powered by the collective energy harvested from the mechanical vibration of the road caused by the passing vehicles and from the temperature gradient between the asphalt of the road and the soil underneath. A stacked piezoelectric transducer converts mechanical vibrations into electrical energy and a thermoelectric generator harvests the thermal energy from the thermal gradient. Electrical energy signals from the stacked piezoelectric transducer and the thermoelectric generators are converted into usable DC power to recharge the battery using AC-DC and DC-DC converters working simultaneously. The multi-sensing system comprises an embedded system with a microcontroller that acquires data from the sensors and sends the sensory data to an IoT transceiver which transmits the data as RF packets to an ethernet gateway. The gateway converts the RF packets into Internet Protocol (IP) packets and sends them to a remote server. Laboratory and road-testing results showed over 98% sensory data accuracy with the system functioning solely powered by the energy harvested from the alternative energy sources. The successful maximum transmission distance obtained between the IoT, and the gateway was approximately 1 mile, which is a considerable transmission distance achieved in an urban environment. Successful operation of the self-powered multi-sensing system under both laboratory and road conditions contributes considerably to the fields of energy harvesting and self-powered remote sensing systems. The energy flow chart and efficiency for the steps in the system were found to be mechanical power from vehicles to the energy harvester of 0.25%, stacked PZT transducer efficiency was found to be 37%, and for the TEGs the efficiency is 11%. AC-to-DC and DC-to-DC converters' efficiencies were found to be 90% and 11%. The wireless communication RF transceiver efficiency was found to be 62.5%.

Keywords: remote sensing; IoT transceiver; gateway; the things network (TTN); stacked PZT transducer; thermoelectric generator (TEG); energy converter

Citation: Dipon, W.; Gamboa, B.; Estrada, M.; Flynn, W.P.; Guo, R.; Bhalla, A. Self-Sustainable IoT-Based Remote Sensing Powered by Energy Harvesting Using Stacked Piezoelectric Transducer and Thermoelectric Generator. *Micromachines* **2023**, *14*, 1428. https://doi.org/10.3390/mi14071428

Academic Editors: Kenji Uchino, Qiongfeng Shi and Huicong Liu

Received: 15 June 2023
Revised: 11 July 2023
Accepted: 11 July 2023
Published: 15 July 2023

1. Introduction

The advent of smart technologies in all aspects of our life has given an incredible boost to the use of remote sensing with real-time data transmission ability and high data accuracy. Remote sensing is in widespread use among various industries, research works, and in our smart homes. Sensing and data acquisition are parts of an essential revolution taking place to automate every walk of our lives. From space exploration to underwater research, every venture is assisted to a great extent by the use of sensing technologies and data acquisition. Remote sensing, in particular, is widely used across research works and industries for acquiring data wirelessly from the sensor. This not only reduces the installation cost by eliminating wires but also allows users to be stationed far from the site of sensing. Remote sensing is the backbone of IoT technology across all applications and, hence, it is an integral part of automation where IoT devices are at the forefront. Most remote sensing that is in use or in the process of being used depends on either batteries or conventional power to run them. Batteries have a lifetime and then need to be disposed of while the devices are replaced with new batteries. This action adversely affects the climate and environment by disposing of toxic materials such as lead. As the widespread use of IoT devices is taking place, more batteries will need to be disposed of, consequently contributing to dire climate and environmental scenarios. The conventional perspective of fighting this climate and environmental effects is to focus on increasing the use of already widely accepted renewable energy sources. These are mainly solar, wind, thermal, and hydropower. Ritchie, Hannah, Max Roser, and Pablo Rosado mentioned this in their paper [1]. Figure 1 shows the energy generation for each category of renewable energy sources until the year 2019 from a survey done by British Petroleum (BP). This indicates the vast difference between hydro and wind energy generation with solar and other renewable sources. The other renewables also consist of biofuel and thermal sources. However, there are many other sources of energy in the surroundings that can be harvested and used for various applications. As the data show, there is not much work done or underway to harvest and use these alternative energy sources which are readily available in the surroundings.

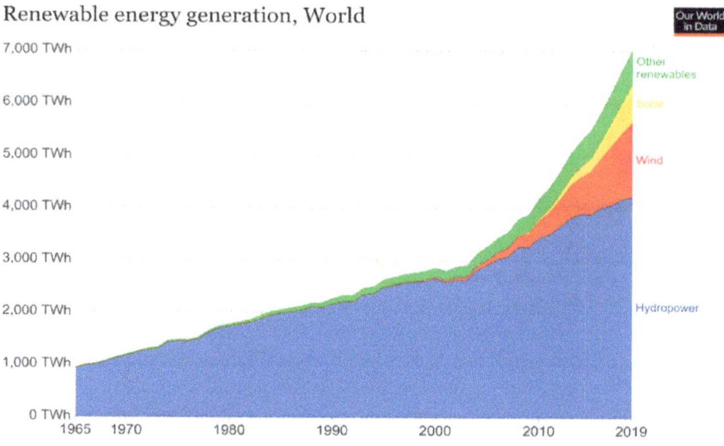

Figure 1. Categories and amount of renewable energy generation (2020) [1].

Lukai Guo and Qing Lu mentioned the concept of collective energy harvesting from pavements using piezoelectricity and thermoelectric generators in their review paper [2]. They stated that most studies on the piezoelectric effect application with piezoelectric transducers (PZTs) showed its limitation in the amount of instantaneous electricity output, while a limited number of studies indicated that a pipe system cooperating with a thermoelectric generator (TEG) may produce more electric power and so has more application potential in energy harvesting pavements. As per their paper, a piezo-thermoelectric (PZT-TEG) system is most effective given the cost–benefit consideration compared to only having PZTs embedded in the pavements. Lallmamode, MA Mujaahiid, and AS Mahdi Al-Obaidi presented a PZT-TEG system to harvest energy from vehicle transportation on highways [3]. However, the electrical energy output from their system, which is 0.2 mW, is not enough to sustain the functional operation of a remote sensing system. Han, Yanhui, Yue Feng, Zejie Yu, Wenzhong Lou, and Huicong Liu also presented a piezoelectric energy harvesting sensor network deployed in a weak vibration environment [4]. The system they presented is very complex with a battery-powered voltage controller. Also, the energy conversion efficiency is reported to be 42%, which is not efficient enough considering the energy conservation requirement for any alternative energy harvesting system. Mysorewala, Muhammad Faizan, Lahouari Cheded, and Abdul Rahman Aliyu also mentioned exploiting hybrid energy harvesting techniques (for example, using wind–solar–thermal energies for outdoor harvesting and radiation vibration for indoor harvesting, with both sets of combined energy sources working either in conjunction with or independently of the normally used batteries) as these combined techniques tend to compensate for one another, thus increasing the likelihood of securing an uninterruptible energy source for the wireless sensor network [5]. Shaikh, Faisal Karim, and Sherali Zeadally, in their comprehensive study of different energy harvester systems for wireless networks, mentioned electrostatic or piezoelectric generators as harvesters for harvesting mechanical vibrations [6]. The piezoelectric generator they stated has an output of 115.2 mW. This generated power is insufficient to provide sustainable energy to a relatively long-range wireless communication remote sensor since communication consumes the most energy of a remote sensing system. This paper presents research that focuses on harvesting mechanical vibrations from roadways and the thermal gradient between the asphalt on the road and the soil underneath. The energies harvested from these sources are then converted into electrical power to run an IoT-based remote multi-sensing system. Transducers are required to convert the mechanical vibrations and thermal gradient into electrical power. In this research, we are using a stacked PZT (lead zirconate titanate) transducer and a combination of thermoelectric generators (TEGs) to convert the mechanical vibrations and thermal gradient, respectively, into electrical energy. An earlier work of the authors used the same stacked PZT transducer to conduct multi-sensing, however, using Bluetooth low energy (BLE) as the communication protocol [7]. Thus, the communication range was limited to approximately 10 feet. This research aims to overcome and resolve some of the shortcomings of contemporary remote sensing systems that are using alternative energy source harvesting to power relatively long-range wireless communication protocols for remote sensing. Improvements in the areas of material engineering of the PZT transducer to the heat sink design are incorporated to produce sufficient energy that can sustain a functional IoT-based remote multi-sensing system.

2. Materials and Methods

2.1. Energy Harvesters: Stacked PZT Transducer and Thermoelectric Generators (TEGs)

The generation of electrical charges in response to applied mechanical stress is known as the direct piezoelectric effect, while the generation of mechanical strain in response to electrical charges is known as the converse piezoelectric effect [8,9]. The electromechanics underlying the direct piezoelectric and converse piezoelectric effects can be best explained with Equations (1) and (2), where x is the mechanical strain [m/m], E is the electric field [V/m], and, in Equation (2), P represents electric displacement [C/m^2] and X is the stress [N/m^2]. In both equations, d is the piezoelectric coefficient [C/N or m/V] which determines the sensitivity of the response of the sample under stress or in an applied electric field [8]. These relationships of piezoelectric materials can be utilized to make piezoelectric materials function as energy harvesters.

$$\text{Direct piezoelectric effect} : P_i = d_{kij}X_{jk} \tag{1}$$

$$\text{Converse Piezoelectric effect} : x_{ij} = d_{ijk}E_k \tag{2}$$

The stacked PZT transducer is custom fabricated from thin plates of PZT-5H-based soft PZT materials. The advantage of using PZT-5H is that it is easily polarized compared to PZT-8, which is a hard PZT material. Figure 2a shows the design of the assembled stacked PZT transducer. The transducer consisted of 21 active plates and one insulating plate. Each PZT plate was of a length and width of 20 mm and a thickness of 2 mm. To ease the alignment issues during fabrication, the stacks were first built in separate 10 and 11 plates and then combined. Indium was chosen as the electrode bonding between the plates and is connected in a way that forms parallel connections between the plates. The final fabricated stacked PZT transducer is shown in Figure 2b, with the dimensions h = 50 mm and l = w = 20 mm [8].

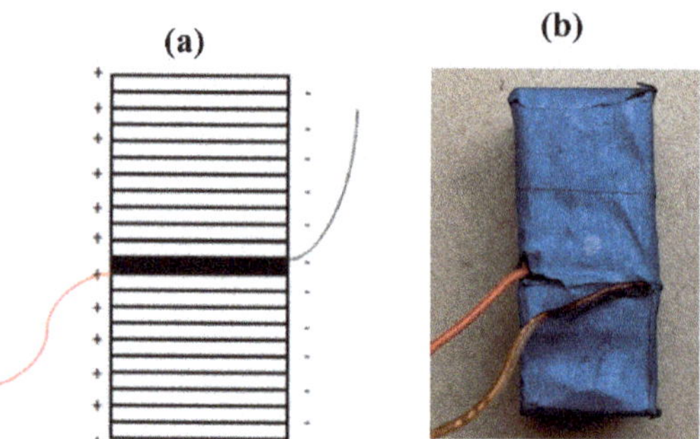

Figure 2. (**a**) Block diagram of the assembled stack, (**b**) final stacked PZT transducer [7].

Gamboa et al. fabricated a stacked PZT transducer for optimized energy harvesting [8] and compared it with various piezoelectric transducers for optimized energy harvesting. Gamboa et al. conducted testing and evaluation of three types of stacked PZT transducers: a 1:3 composite stacked PZT transducer, a specially designed and fabricated stacked PZT transducer, and a commercially available stacked sample. PZT-5H samples were used to assemble the stacked PZT transducer. Piezoelectric material PZT-5H is used to harvest energy in wireless sensor networks due to its very high permittivity and sensitivity properties. These materials have electromechanical properties which are used to harvest energy [9]. The fabricated transducer gave maximum power density per unit of transducer volume, measured at 0.615 mW/mm^3 at 965 KN/m^2 (140 psi). Considering power density per unit of transducer volume as the more appropriate way of determining the type of stacked transducer to be used, the fabricated stacked PZT transducer was chosen to be used for this research. The PZT plates are mechanically in series and electrically in parallel to increase the current from the stacked PZT transducer.

Thermoelectric generators (TEG) are solid-state semiconductor devices that convert a temperature difference and heat flow into a useful DC power source. Thermoelectric generator semiconductor devices utilize the Seebeck effect to generate voltage. This generated voltage drives electrical current and produces useful power at a load. The Seebeck effect is the generation of electricity between a thermocouple when the ends are subjected to the temperature difference between them. A combination of four thermoelectric generators is used in the research. Bismuth telluride (Bi$_2$Te$_3$) TEGs were selected due to the high figure of merit. The figure of merit is the number used to determine the efficiency of a TEG to convert a temperature difference into useful electrical energy. The figure of merit for different materials varies with temperature. Considering the system will be deployed in Texas, where maximum temperature difference occurs during summertime causing the asphalt temperature to reach 400 K, bismuth telluride TEGs have the highest figure of merit at this temperature. Figure 3a illustrates the Seebeck effect considering the temperature gradient between the asphalt and the surrounding soil. The heat flows from the asphalt to the soil underneath. Heat sinks are used between the top and bottom plates to carry the heat from the asphalt to the top metal junction of the TEGs. The bottom metal junction of the TEGs is at the same temperature as the soil underneath the asphalt. Figure 3b shows the COMSOL model of the heat sink concept as a module with the TEGs as well as isolated. Figure 3c shows the fabricated heat sink module along with TEGs inside. The heatsink design was validated by COMSOL simulation, and the model indicated the TEG was able to increase the temperature difference across the TEG from 0.1 °C to 1.1 °C [10]. The TEGs were tested with the module in direct sunlight. A total of 4 TEGs are divided into 2 groups, each comprising 2 TEGs connected in parallel are in series with the other group, making 4 TEGs in total. This combination is done to attain optimized voltage and current from the TEGs. The stacked PZT transducer has an energy conversion efficiency of 37% and each TEG has an efficiency of 11%.

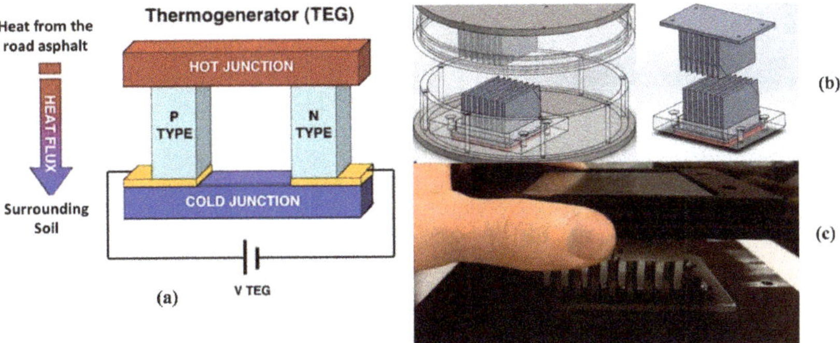

Figure 3. (**a**) Illustration of the Seebeck effect. (**b**) COMSOL model of the heat sink in the module along with heat sink isolated. (**c**) The fabricated TEG and heat sink module [10].

2.2. Converters for Stacked PZT Transducer and TEGs

The electrical signals from the stacked PZT transducer and the TEGs are sinusoidal and DC, respectively. Hence, it is required to have a combination of AC-to-DC and DC-to-DC boost conversions in the system. The voltage from the stacked PZT transducer is very high and, hence, needs to be reduced. The voltage from the TEGs is low and, hence, is required to be boosted to the charging voltage of the battery. The AC-to-DC conversion aims at collecting as much charge as possible from the PZT. As per the maximum power transfer theorem, maximum power transfer from the source to the load is attained when the impedances of the source and the load match. The stacked PZT transducer can be represented as an electrical circuit consisting of a current source, a resistance, and a capacitor all in parallel. The electrically parallel combination of the PZT plates in the stacked PZT transducer creates an increase in the capacitive impedance of the circuit which leads to lower total impedance of the source. This is an important feature as it facilitates impedance matching with the impedance of the AC-to-DC converter, which is typically lower. Consequently, higher charge collection is made possible due to the impedance matching. The AC-to-DC converter also abruptly and efficiently reduces the high voltage generated by the stacked PZT transducer by rapidly collecting the charges once they are generated. Since the transducer acts as a capacitor, its voltage is a function of the charges accumulated in the transducer. Once these charges are extracted from the transducer to charge the battery, the high voltage generated by the transducer drops abruptly. Hence, a capacitor-based converter is selected since a capacitor acts as a short circuit at the start. Consequently, the generated charges are collected immediately to drop the voltage of the transducer. An EH301A (Sourced from Advanced Linear Devices Inc., Sunnyvale, CA, USA) energy harvester, shown in Figure 4a, is used as the AC-to-DC converter for the stacked PZT transducer. It is equipped with a 3300 µF storage capacitor which is charged by the charges from the stacked PZT transducer. The output voltage rises until it reaches the charging voltage of the battery when the accumulated charges are transferred into the battery until the output voltage drops below the charging voltage and the cycle repeats. This cycle and different stages are shown in Figure 4b, where V_P represents the output voltage of the converter which switches between V_H and V_L representing high and low voltage levels. The conversion efficiency of the AC-to-DC converter was found to be 90%.

The voltage from the TEGs, which is 120 mV, is very low and needs to be boosted to the charging level of the battery. ELC-BVB120 boost convert is selected for the TEGs because its conversion efficiency was experimentally found to be 86% while converting 120 mV to about 4 V with a load. The objective was to have both AC-to-DC and DC-to-DC boost converters operate simultaneously, charging the rechargeable battery pack of NiCad batteries. The Pspice model for the combined converter design is shown in Figure 5. It was recorded that multi-sourcing energy is very possible by the carefully designed electronic module. Minimum energy sourced from harvesters that can be converted to collect energy cumulatively to charge a battery thus also proved to have high efficiency of the module. Four TEGs are connected to the DC-rail using the concept that was presented before. Current responses are equivalent to what one should expect from a TEG at the 10 V (open circuit) output of the boost converter. A low thermal variant, that is, a small temperature difference across the TEG plates, is considered in this simulation because the soil is an excellent thermal conductor and, hence, the temperature difference is assumed to be small between the asphalt on the road and the soil underneath. To make more rigorous verification of the design, TEGs were given some uneven internal impedance to create non-uniform current responses. All resistors are used in line with the energy source of devices to control current and make them equivalent to the actual test results obtained from the boost converters. Since boost converters were not incorporated into the schematic, equivalent circuits were designed in the schematic. All multi-sources at PZTs' AC-rail and TEGs' DC-rail are acting to charge the battery flawlessly. Continuously feeding current is possible by this technique, because a rechargeable battery is a very large capacitor in real life. The real-time tested power was observed to be more than expected, with an improvement of 30.27% in efficiency. Inductive and capacitive converter combinations might have produced better impedance responses, resulting in a higher current to charge the battery. Figure 6 shows the stacked PZT transducer, TEG heatsink modules, and the converters inside the mechanical enclosure called HiSEC (Hybrid Integrated Sensing and Energy Conversion) module.

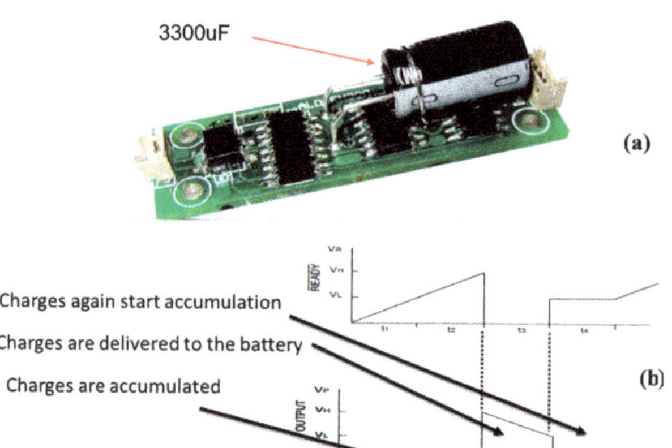

Figure 4. (**a**) EH301A energy harvester with 3300 uF storage capacitor. (**b**) Charging and discharging cycle of the EH301A energy harvester.

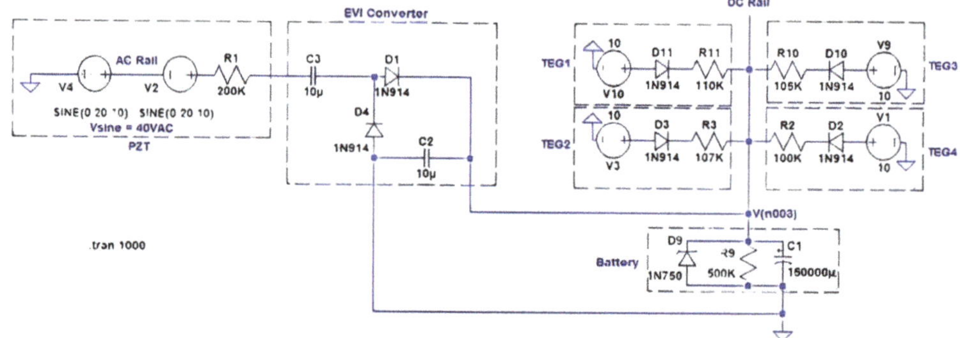

Figure 5. Pspice model design of an equivalent circuit of the system for verification.

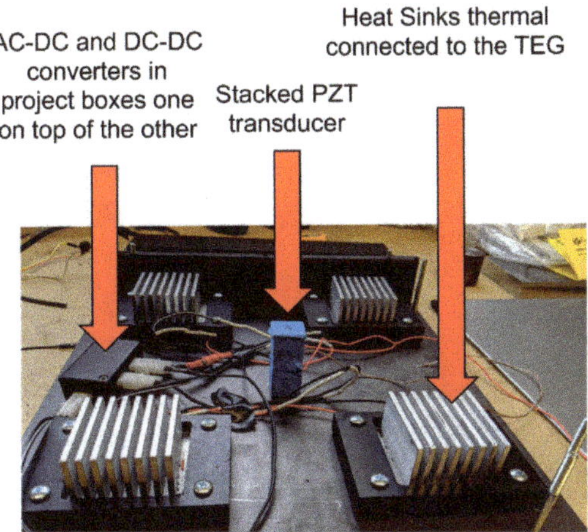

Figure 6. Hybrid Integrated Sensing and Energy Conversion (HiSEC) module with the stacked PZT transducer, heat sinks for the TEGs, and the converters.

2.3. Sensor Hardware, Algorithm, and Wireless Communication

Three sensing parameters are transmitted over wireless communication. The system is developed to be deployed buried under the roadway as it is intended to harvest energy from the vehicles passing over the road and the thermal gradient between the asphalt and the soil underneath. A PZT-5H based 2 mm × 2 mm × 2 mm piezoelectric plate was used as the piezoelectric-based pressure sensor to estimate the weight of the vehicle. The PZT plate also acts as the sensor counting the axle counts of the vehicles. Two negative temperature coefficient (NTC) temperature sensors were used to read the temperature difference between the asphalt and the soil. Figure 7a shows the piezoelectric pressure sensor and Figure 7b shows the NTC based temperature sensors installed on the heat sink and the lower plate of HiSEC. A microcontroller with an algorithm coded does all the computation required to estimate the sensing parameter values. The microcontroller used in the research is a 16 MHz ATmega328P-based microcontroller called Arduino Nano (Sourced from Microchip Technology Inc., Chandler, AZ, USA) which uses RISC architecture with an ADC resolution of 10 bits. It has a flash memory of 32 KB, of which 2 KB is occupied by the bootloader. The other smaller option was Arduino Pro Mini, which also has lower power consumption. However, Arduino Pro Mini does not come with a USB port and, hence, it is difficult to load programs onto it. Also, Arduino Nano does not occupy much space and the slightly higher power consumption compared to Arduino Pro Mini is compensated for by the ease of programming and reprogramming if required.

The communication protocol used in the research is LoRa WAN with an operating frequency of 915 MHz and the default server used to collect and display the sensor data is called The Things Stack or TTN server, which is a LoRa WAN network server. The LoRa WAN protocol uses Chirp Spread Spectrum (CSS) modulation technology. The hardware of the LoRa WAN consists of an IoT transceiver with a patch antenna fabricated on it which can be extended into a dipole antenna by soldering it. An ethernet gateway is connected to the internet through an ethernet connection. The IoT transceiver receives serial data from the microcontroller of the embedded multi-sensing system using a Serial Peripheral Interface (SPI) protocol embedded in the codes. The serial data are then converted into radio frequency (RF) packets and sent wirelessly to the nearby The Things Network gateway which converts these RF packets into Internet Protocol (IP) packets and transmits them to The Things Network (TTN) server as a stream of hexadecimal values. Since the hexadecimal version of the data—which is the default data type on the TTN server—is not user-friendly, a console server is established to display the sensory data in a user-friendly manner. This was done by a Java script and a free server account. This script takes the hexadecimal data from the TTN server and segregates them under the corresponding sensor category and displays them as decimal values on the console server. The hardware of the LoRa WAN and the entire communication steps are shown in Figure 8. The completed prototype of the HiSEC is shown in Figure 9.

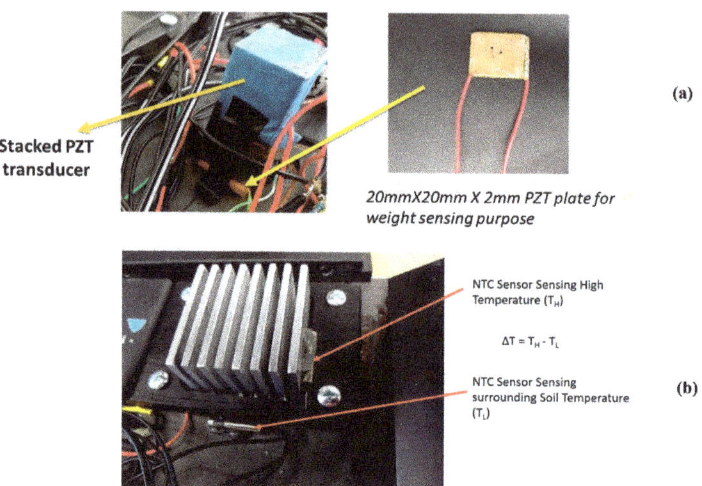

Figure 7. (**a**) Piezoelectric transducer-based tire pressure sensing. (**b**) Temperature difference measurement using NTC temperature sensors.

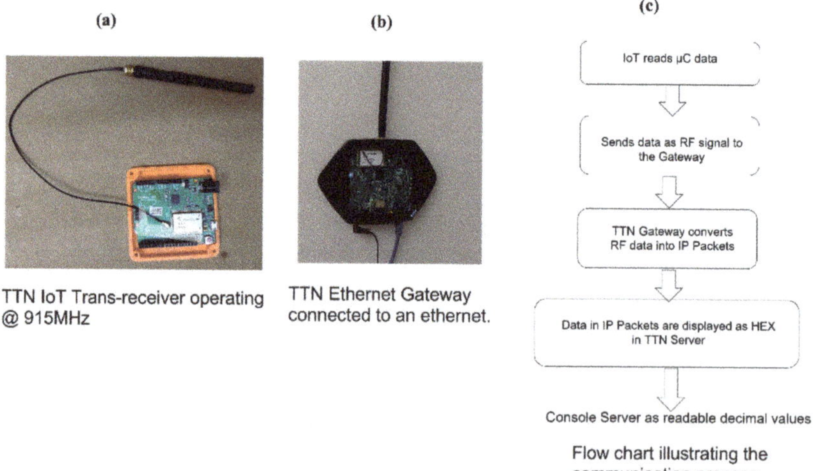

Figure 8. (**a**) TTN IoT transceiver. (**b**) TTN ethernet gateway. (**c**) Flow chart illustrating the communication steps.

Figure 9. A prototype demonstration unit of HiSEC with main components labeled.

3. Results and Discussion

An MTS Acumen electrodynamic testing platform is used to test the system in a laboratory environment under various forces and frequencies (Figure 10). These forces are used to test the electrical power conversion from mechanical vibrations of the stacked PZT transducer. Real-time sensory data in hexadecimal forms and decimal forms were displayed on the TTN server and the console server, respectively. Power measurements from the stacked PZT transducer and the TEGs were also recorded under various forces and temperature differences. Several forces at different frequencies are used to test the HiSEC; also, temperatures were varied over a range to replicate the temperature gradient expected under roads in Texas. Two sensory parameters were recorded during the tests to prove the concept, vehicle weight, and axle count (represented as vehicle count). Figure 11 shows the real-time sensory data displayed on the TTN server and the console server when the MST Acumen ran with 1.5 KN of force at 10 Hz for 60 s. Similar data for 2 KN of force at 10 Hz for 60 s are shown in Figure 12. These equal 101.6 kg (224 lbs.) and 203.2 kg (448 lbs.) of forces at 10 Hz for 60 s, respectively.

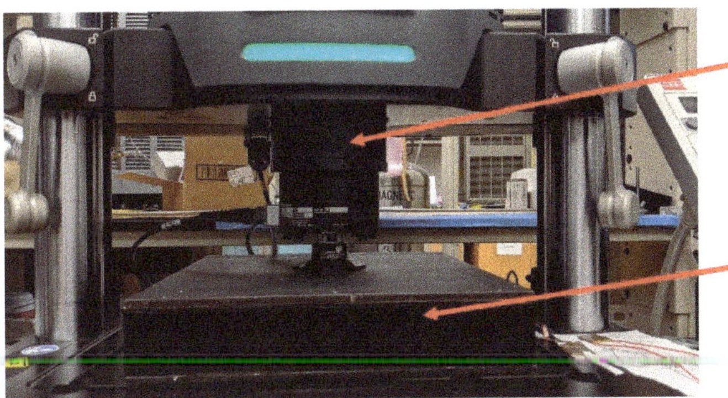

Figure 10. HiSEC prototype under test using MTS Acumen electrodynamic testing platform.

APPLICATION DATA

Filters	uplink	downlink	activation	ack	error

	time	counter	port	
▲	14:28:36	48	1	payload: 00 00 E0 3A 09 00 00 00 00 00 00 00 00 00 00 00 00 00 C
▲	14:28:33	47	1	payload: 00 00 E0 3A 09 00 00 00 00 00 00 00 00 00 00 00 00 00 C
▲	14:28:29	46	1	payload: 00 00 E0 3A 09 00 00 00 00 00 00 00 00 00 00 00 00 00 C

Console

Main Page

Show: 50   Search:

Count ▼	Date	Time	Port	Payload	Vehicle Count
97	04/02/2020	14:31:17	1	224	589
96	04/02/2020	14:31:14	1	224	589
94	04/02/2020	14:31:07	1	224	589
93	04/02/2020	14:31:04	1	224	589
92	04/02/2020	14:31:00	1	224	589

Figure 11. Sensory data (1.5 KN, 10 Hz, for 60 s) displayed in hexadecimal format on the TTN server and converted into a decimal format to be displayed on the console server by the Java script.

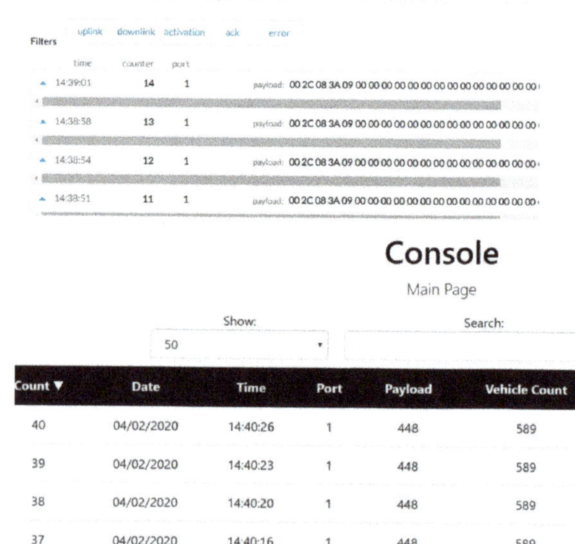

APPLICATION DATA

Filters	uplink	downlink	activation	ack	error

	time	counter	port	
▲	14:39:01	14	1	payload: 00 2C 08 3A 09 00 00 00 00 00 00 00 00 00 00 00 00 00
▲	14:38:58	13	1	payload: 00 2C 08 3A 09 00 00 00 00 00 00 00 00 00 00 00 00 00
▲	14:38:54	12	1	payload: 00 2C 08 3A 09 00 00 00 00 00 00 00 00 00 00 00 00 00
▲	14:38:51	11	1	payload: 00 2C 08 3A 09 00 00 00 00 00 00 00 00 00 00 00 00 00

Console

Main Page

Show: 50   Search:

Count ▼	Date	Time	Port	Payload	Vehicle Count
40	04/02/2020	14:40:26	1	448	589
39	04/02/2020	14:40:23	1	448	589
38	04/02/2020	14:40:20	1	448	589
37	04/02/2020	14:40:16	1	448	589

Figure 12. Sensory data (2 KN, 10 Hz, for 60 s) displayed in hexadecimal format on the TTN server and converted into a decimal format to be displayed on the console server by the Java script.

Similar testing of the system was also conducted on the roadways with successful results. A wooden ramp was used to drive cars over it while the HiSEC was in place on the ramp, as shown in Figure 13. The maximum communication range is over a distance of approximately 1.12 km in an urban environment. However, the range was expected (more than 1 mile) to be farther than achieved without an intermittent drop in the communication channel operating at 915 MHz. Path loss will represent the amount of energy lost in free space over a distance between the transmitter and the receiver. The farther away Tx is from Rx, the lower the energy is. Path loss is usually expressed as shown in Equation (2):

$$\text{FSPL} = \left(\frac{4\pi d}{\lambda}\right) \times 2 = \left(\frac{4\pi d f}{c}\right) \times 2 \tag{3}$$

where FSPL = Free Space Path Loss, d = distance between Tx and Rx in meters, and f = frequency in Hertz. There is also a widely used logarithmic formula for free space attenuation as shown in Equation (3):

$$\text{FSPL(dB)} = 20\log_{10} d + 20\log_{10} f - 147.55 \tag{4}$$

Figure 13. Road testing of the HiSEC module with the TTN ethernet gateway placed 30 m from the ramp.

The transmitter and the receiver are located 1.12 Km, which is 1120 m when the system was tested outside. Putting d = 1120 m in the above formula, the FSPL comes out to be 27.34 dB.

COMSOL modeling and simulation of the dipole antenna between two metal plates were done to find the reasons for the shorter communication range and unstainable communication. It was concluded that the electromagnetic waves from the antenna cause eddy currents to develop at the metal plates, which inhibit the transmission of the radio frequency signals to the ethernet gateway. The COMSOL simulation result is shown in Figure 14.

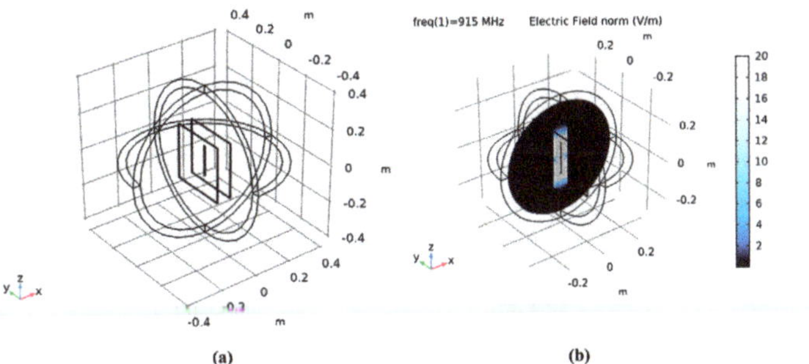

Figure 14. (**a**) COMSOL model of a dipole antenna between two metal plates. (**b**) COMSOL model simulation showing electric fields developing at the plates' surfaces due to eddy currents.

The sensory data attained 98% in data accuracy for dynamic sensing of vehicle weights and axle counts, which is significantly higher than contemporary systems and conventional weight-in-motion (WIM) systems. Experimentally, the power consumption of the IoT transceiver was found to be 40 mW for 15 s each time it transmits data. That is, 0.6 W power is required for every transmission. The batteries used in the system have 800 mA-h at 3.6 V, equal to 2880 mW per hour, which is sufficient enough to restore operation during a lack of energy from the transducer. For estimating the energy that will be generated by the stacked PZT transducer and the TEGs, we considered the US-59 highway as an example. Table 1 provides the traffic data of highway US-59 by the Texas Department of Transport (TxDOT). Here, ADT is average daily traffic and ADDT is average daily truck traffic. It also gives the average truck speed, which is 29.73 ms^{-1}. The stacked PZT transducer generated on average approximately 0.8 mW of power at 140 PSI. From the traffic data and the experimentally found power generation data, it can be stated that the stacked PZT transducer will generate approximately 4 W of power per day while the TEGs will generate on average 1 W of power per day. Thus, approximately eight successful transmissions can be done using the energy generated collectively from the stacked PZT transducer and the TEGs. This consumption can be reduced by increasing the period between each wireless transmission. Hence, it can be estimated that the power generation will be sufficient enough to replenish the battery power consumed by the IoT transceiver.

Table 1. US-59 traffic data collected (TxDOT).

#	Parameter	Value
1	Speed Limit (mph)	75
2	Average truck speed (mph)	66.5
3	Design lane (outside lane) ADT	4208
4	Design lane (outside lane) ADTT	1599
5	Tire pressure (PSI) (vary from truck to car)	30–120

From the electromechanical testing of the HiSEC, it was found that the mechanical energy input from the crosshead of the testing platform to the HiSEC was 3.15 W and the 8mW energy eventually reached the stacked PZT transducer, resulting in 0.25% energy conversion efficiency. This is aligned with our estimate of the roadway scenario since it is estimated that only a small amount of the mechanical energy will reach through the asphalt, grave, soil, and then to the stacked PZT transducer. The LoRa WAN IoT transceiver output power is limited to 25 mW, resulting in an electrical-power-to-signal-power conversion efficiency of 62.5%. This research was intended to prove the concept and efficacy of an IoT-based multi-sensing system, and the efficiency of the developed system is measured in terms of the successful functionality of the system powered exclusively by the energy harvested from the stacked PZT transducer and the thermoelectric generators. The experimental results show that the system is capable of operating successfully with the harvested energy by the energy harvesters. The HiSEC module, which is an integrated system housing the energy harvesters, converters, sensors, and embedded data acquisition circuit, and the IoT transceiver is 12-inch X 12-inch X 2-inch in dimension. However, the module can be redesigned with smaller dimensions for formfitting any particular application. The application of this system can be in the fields of industries to research works where low-power remote sensing plays an essential role and where there is scarcity and inconvenience to the access of conventional power. This system can also play a part in increasing the life span of the batteries of IoT devices and, hence, will positively impact the environment. The system and the processes utilized in the HiSEC can be divided into energy domains, and each domain has its energy conversion efficiency.

4. Conclusions

The work presented in this paper aimed to resolve some of the most critical difficulties involving the widespread use of alternative energy sources as power sources for wireless communications in remote sensing. The HiSEC module is powered by a stacked PZT transducer acting as a mechanical energy harvester and TEGs harvesting thermal gradient. High accuracy of the sensory data was achieved for road traffic axle count, vehicle weight, and temperature difference sense. The system resolves the issues mentioned in contemporary literature by not only achieving high sensing accuracy but also ensuring sustainability in remote sensing data transmission. In comparison to the harvesters for wireless networks mentioned in the literature, which generate power in the microwatt range, the harvester in this research generates power in the range which is sufficient enough to sustain wireless communication for remote sensing. Hence, the power generated is also a significant improvement from contemporary harvesters. However, the standout factor for the developed system in comparison to the state-of-the-art is the energy-efficient way of utilizing the generated power to successfully conduct multi-sensing and wireless communication. The entire system is programmable and additional sensing capabilities can be incorporated with ease. This facilitates the adaptability of the system for multiple applications. The proof of work of this research can be utilized in many ways for other sensing applications powered by alternative energy sources. Through this research and its impact, we envisage an entirely new generation of IoT-based remote sensing based on the concept of being powered by the energy harvested from various alternative energy sources, leading to a positive impact on the environment and significantly extending battery life span. Future improvements include resolving the communication issue with an antenna between the metal plates of the HiSEC and also improving the electrical power output from the harvesters.

Author Contributions: Conceptualization, methodology, software, validation, investigation, data curation, writing—original draft preparation, W.D., B.G., M.E. and W.P.F.; investigation, writing—review and editing, B.G., M.E. and W.P.F.; supervision, project administration, R.G. and A.B.; funding acquisition, R.G. and A.B. All authors have read and agreed to the published version of the manuscript.

Funding: This research has been supported by the Department of Energy National Nuclear Security Administration under grant DE-NA0004003.

Data Availability Statement: All data that are relevant to this paper are provided as images.

Acknowledgments: The authors would like to thank Sean Garnsey, George Nall, and Melinda Duong for their technical contributions.

Conflicts of Interest: The authors declare no conflict of interest.

References

1. Ritchie, H.; Roser, M.; Rosado, P. Energy. Our World in Data. 2022. Available online: https://ourworldindata.org/co2-emissions (accessed on 10 July 2023).
2. Guo, L.; Lu, Q. Potentials of piezoelectric and thermoelectric technologies for harvesting energy from pavements. *Renew. Sustain. Energy Rev.* **2017**, *72*, 761–773. [CrossRef]
3. Lallmamode, M.A.M.; Al-Obaidi, A.S.M. Harvesting energy from vehicle transportation on highways using piezoelectric and thermoelectric technologies. *J. Phys. Conf. Ser.* **2021**, *2120*, 012016. [CrossRef]
4. Han, Y.; Feng, Y.; Yu, Z.; Lou, W.; Liu, H. A study on piezoelectric energy-harvesting wireless sensor networks deployed in a weak vibration environment. *IEEE Sens.* **2017**, *17*, 6770–6777. [CrossRef]
5. Mysorewala, M.F.; Cheded, L.; Aliyu, A. Review of energy harvesting techniques in wireless sensor-based pipeline monitoring networks. *Renew. Sustain. Energy Rev.* **2022**, *157*, 1120.
6. Shaikh, F.K.; Zeadally, S. Energy harvesting in wireless sensor networks: A comprehensive review. *Renew. Sustain. Energy Rev.* **2016**, *55*, 1041–1054. [CrossRef]
7. Dipon, W.; Gamboa, B.; Guo, R.; Bhalla, A. Energy harvesting using a stacked PZT transducer for self-sustainable remote multi-sensing and data logging system. *J. Compos. Sci.* **2022**, *6*, 49. [CrossRef]
8. Gamboa, B.M.; Guo, R.; Bhalla, A. Piezoelectric stacked transducer evaluation and comparison for optimized energy harvesting. *Ferroelectrics* **2018**, *535*, 8–17. [CrossRef]

9. Kalinin, S.V.; Mirman, B.; Karapetian, E. Relationship between direct and converse piezoelectric effect in a nanoscale electromechanical contact. *Phys. Rev. B* **2007**, *76*, 212102. [CrossRef]
10. Gamboa, B.; Estrada, M.; Djikeng, A.; Nsek, D.; Binzaid, S.; Dessouky, S.; Bhalla, A.; Guo, R. Integrated Piezoelectric and Thermoelectric Sensing and Energy Conversion. *Adv. Ceram. Environ. Funct. Struct. Energy Appl. II* **2019**, *266*, 15.

Article

A Piezoelectric MEMS Speaker with a Combined Function of a Silent Alarm

Qi Wang [1,2], Tao Ruan [1,2], Qingda Xu [1,2], Zhiyong Hu [1,2], Bin Yang [1], Minmin You [1], Zude Lin [1,*] and Jingquan Liu [1,*]

1   National Key Laboratory of Science and Technology on Micro/Nano Fabrication, Shanghai Jiao Tong University, Shanghai 200240, China
2   Department of Micro/Nano-Electronics, Collaborative Innovation Center of IFSA, Shanghai Jiao Tong University, Shanghai 200240, China
*   Correspondence: linzude@sjtu.edu.cn (Z.L.); jqliu@sjtu.edu.cn (J.L.)

Abstract: To explore the versatility of speakers, a piezoelectric micro-electro-mechanical system (MEMS) speaker combining the function of a silent alarm is proposed, which mainly comprises a lead zirconate titanate (PZT) actuation layer and a rigid–flexible coupling supporting layer. Measurements performed on encapsulated prototypes mounted to an artificial ear simulator have revealed that, compared to a speaker with a rigid supporting layer, the sound pressure level (SPL) of the proposed piezoelectric MEMS speaker with a rigid–flexible coupling supporting layer is significantly higher and is especially higher by 4.1–20.1 dB in the frequency range from 20 Hz to 4.2 kHz, indicating that the rigid–flexible coupling supporting layer can improve the SPL significantly in low frequency. Moreover, the spectral distribution characteristic of its playback audio is similar to that of the commercial electromagnetic type. The device can also function as a silent alarm based on oral airflows in dangerous situations, as it performs well at recognizing words according to their unique voltage-signal characteristics, and can avoid the effects of external sound noise, body movement, long distance, and occlusion. This strategy provides inspiration for functional diversification of piezoelectric MEMS speakers.

Keywords: piezoelectric; micro-electro-mechanical system (MEMS); rigid–flexible-coupling; speaker; silent alarm

Citation: Wang, Q.; Ruan, T.; Xu, Q.; Hu, Z.; Yang, B.; You, M.; Lin, Z.; Liu, J. A Piezoelectric MEMS Speaker with a Combined Function of a Silent Alarm. *Micromachines* **2023**, *14*, 702. https://doi.org/10.3390/mi14030702

Academic Editor: Fabio Di Pietrantonio

Received: 16 February 2023
Revised: 16 March 2023
Accepted: 20 March 2023
Published: 22 March 2023

1. Introduction

Over recent years, there has been much research interest in audio electronics, such as micro speakers [1,2], hearing aids [3,4], and speech recognition sensors [5–7]. These single-function devices have greatly improved people's quality of life, but they are still far from meeting the ever-increasing demands for convenience. Therefore, to avoid the inevitable heaviness and messiness caused by wearing various single-function devices, dual function or multifunction wearable devices have become desirable and constitute new research trends [8–12]. In particular, the rapid development of micro-electro-mechanical system (MEMS) technology is expected to realize the low-cost mass production of such devices.

Existing speakers are mainly divided into electromagnetic, electrostatic, and piezoelectric types according to the working mechanism. The traditional electromagnetic speaker based on electromagnetic induction still cannot be replaced by the MEMS one because of its relatively large permanent magnet [13,14]. Despite offering benefits such as low membrane mass and ease of integration, electrostatically-driven capacitive MEMS speakers are still subject to limitations such as the pull-in effect and the requirement for high driving voltage, as highlighted in [15,16]. Piezoelectric MEMS speakers based on the reverse piezoelectric effect are gradually attracting attention for the advantages of low power consumption, high theoretical sound pressure level (SPL), fast response, as well as being dust-free, and viable for mass production. Therefore, piezoelectric materials provide a potential solution for MEMS speakers [17–19]. Researchers have developed various piezoelectric materials, such as ZnO [20,21], AlN [22], PZT [23,24], PMN-PT [25–27], and PZN-PT [28]. The present choice of sputtering piezoelectric PZT thin film as the preferred material is related to not only our material preparation process but also the fact that PZT has the advantages of higher piezoelectric charge constant and electromechanical coupling coefficient [29–31].

Although researchers have adopted new working principles, structures, and driving electrodes [2,29,32,33], piezoelectric MEMS speakers still suffer from the challenge of low SPL, and therefore cannot realize multifunction. In our previous work, the obtained SPL is still small when the radius is 1.5 mm [32], which needs to be further considered to reduce structural stiffness. In addition, existing speech-recognition devices are either disturbed by external sound noise [34], body movement [35], or are limited by distance or occlusion [6,7], thereby leading to errors or failures in speech recognition. These problems have been encouraging researchers to find a more accurate and convenient way of the silent speech recognition system. This work would gain further significance if the aforementioned piezoelectric MEMS speaker, which currently serves a single function, could also function as a silent alarm.

Herein, a piezoelectric MEMS speaker combining function of silent alarm is demonstrated. The deposition of 10 μm thick parylene C on the upper surface before etching the back cavity is the key process when forming the rigid–flexible coupling supporting layer. Moreso, the rigid vibration membrane with a maximum radius of 2 mm is a central circle surrounded by concentric double rings. Compared to a speaker with a rigid supporting layer, this proposed piezoelectric MEMS speaker with rigid–flexible coupling supporting layer can improve the SPL significantly in the frequency range from 20 Hz to 4.2 kHz. Moreover, in the frequency range of 20 Hz to 20 kHz, a volume higher than 72.3 dB SPL is achieved at driving voltage of 10 V, which satisfies the basic needs of human hearing. The proposed device can also function as a silent alarm based on oral airflow, which performs well at identifying an alarm silently through its unique voltage-signal characteristics, and avoids the effects of external sound noise, body movement, long distance, and occlusion. This proposed device provides inspiration for expanding the capabilities of piezoelectric MEMS speakers.

2. Strategy and Design

The strategy and design of the piezoelectric MEMS speaker combining function of silent alarm are shown in Figure 1a,b. It comprises a vibration membrane with a central circle surrounded by concentric double rings and a flexible parylene C supporting layer. The shape and rigid–flexible coupling structure of the vibration membrane were designed to reduce structural stiffness, which can increase the vibration displacement, thereby facilitating the increase in SPL [32] or voltage signal for a given oral airflow blowing onto the surfaces [36]. Moreover, the parylene C was deposited on the device surface and back cavity to form the rigid–flexible coupling supporting layer and for avoidance of the influence of water vapor on its performance. The proposed device can function as a speaker by using the inverse piezoelectric effect of the PZT thin film and can also perform well at recognizing words according to their unique voltage-signal characteristics when spoken silently. As shown in Figure 1c, the size of the chip is less than 10 mm, and the shape of the vibration membrane can be clearly displayed. The functions of the proposed device are shown in Figure 1d.

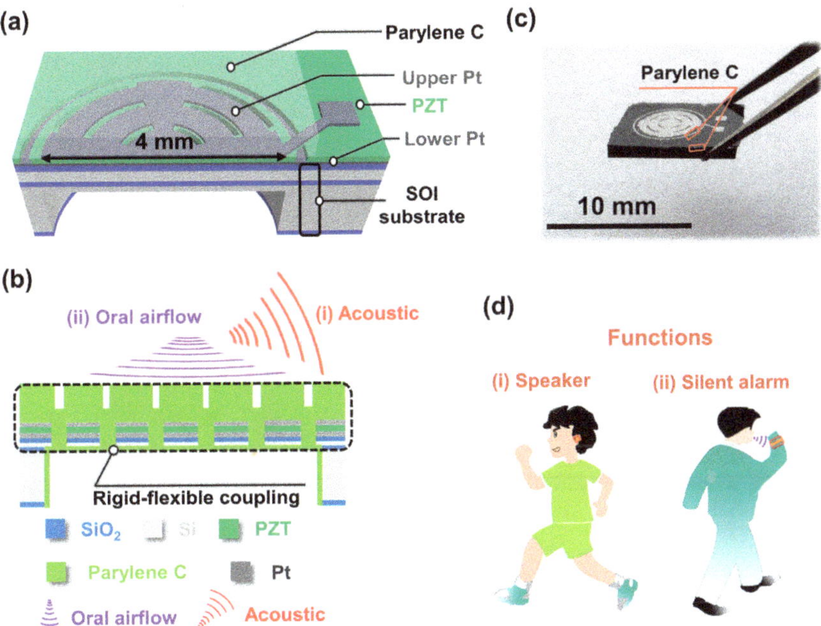

Figure 1. Strategy and design of piezoelectric MEMS speaker combining function of silent alarm. (**a,b**) Strategy and design, (i) Acoustic, (ii) Oral airflow. (**c**) Photographs of chip clamped by tweezers. (**d**) Piezoelectric MEMS speaker (i) combining function of silent alarm (ii).

3. Materials and Methods

The preparation process of the chip is shown in Figure 2a. The thickness of the buried oxygen layer SiO_2 was ~1.5 µm, the thickness of the top Si layer on the SOI (silicon on insulator) wafer was ~2 µm, and ~500-nm-thick SiO_2 was deposited on the top Si surface by plasma-enhanced chemical vapor deposition. Finally, a ~100-nm-thick Pt layer, a ~1-µm-thick PZT piezoelectric thin film was deposited by RF magnetron sputtering (piezoelectric coefficient (d_{31}) = ~−220 pm/V, dielectric constant (ε_r) = ~900, and Young's modulus = ~70 × 10^9 Pa), and a ~100-nm-thick Pt layer were sputtered successively on the deposited SiO_2 surface (Figure 2(a1)). The sputtered PZT does not require a dedicated polarization due to the self-polarization effect.

Then, the upper electrode Pt layer was etched to form a central circle surrounded by concentric double rings (Figure 2(a2)) (see Figure S1 in the Supplementary Materials for detailed sizes), then BHF was prepared in the volume ratio of V[NH_4F(40%)]: V[HF] = 1:5, and the etching solution was obtained by mixing a solution with a volume ratio of V[BHF]: V[HCl]: V[H_2O] = 1:25:175. The etching residue attached to the PZT surface was removed after etching the 1-µm-thick PZT ((Figure 2(a3)) [37,38]. The process of etching the lower electrode was the same as that for the upper electrode (Figure 2(a4)). The deposited SiO_2, top Si, middle buried oxygen layer SiO_2 were then etched in sequence (Figure 2(a5)). Then, ~10 µm parylene C was deposited on the membrane surface (Figure 2(a6)). Then, the back cavity was formed after etching barrier SiO_2 (Figure 2(a7)). Finally, a ~0.5-µm-thick layer of parylene C was deposited on the device surface and back cavity to avoid the influence of humidity on device performance by isolating water vapor (Figure 2(a8)) [39–41]. Of course, the functions of speaker and silent alarm are exclusive, so that a circuit rotation device and a wireless data transmission module need to be built to facilitate the conversion of functions and data transmission. However, these efforts are still in the research process due to the current limited knowledge in this area.

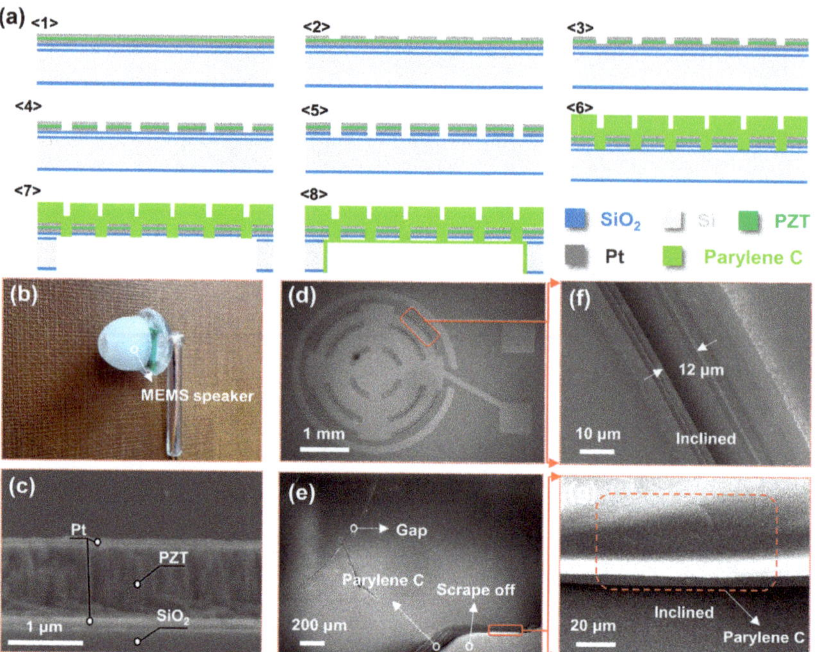

Figure 2. Preparation process and SEM images of the piezoelectric MEMS speaker combining function of silent alarm. (**a**) Process of fabricating the proposed device: (1) materials, (2) etching the upper electrode, (3) wet etching the PZT, (4) etching the lower electrode, (5) etching gaps on the upper surface, (6) depositing thick parylene C, (7) etching the back cavity, (8) depositing thin parylene C. (**b**) Encapsulated speaker. Scanning electron microscopy (SEM) images of (**c**) PZT, (**d**) the whole surface of the device without parylene C, and (**e**) the deposited parylene C, and (**f**) the inclined gap, and (**g**) the inclined view of parylene C section.

The SEM images were characterized by a scanning electron microscope (FE-SEM, ZEISS Gemini, Oberkochen, Germany). Impedance Analyzer (KEYSIGHT, E4990A, Santa Rosa, CA, USA) was used to measure the relationship between the frequency and impedance or phase angle, respectively. The SPLs of the piezoelectric MEMS speakers were characterized by an audio analyzer (FX100, NTi Audio, Schaan (Liechtenstein), Switzerland). The SPLs distributions at each moment were recorded by an audio and acoustic analyzer (XL2, NTi Audio, Schaan (Liechtenstein), Switzerland). The acoustic tests were all carried out in an anechoic box. The output voltage signals and voltage-signal characteristics of the device were recorded by a digital storage oscilloscope (DSO-X 2024A, California, Agilent). Parylene C was deposited by using a parylene C thin-film deposition system (PDS 2010; SCS, Indianapolis, IN, USA).

4. Discussions

4.1. Photograph and SEM Images

Figure 2b shows the device encapsulated with a 3D-printed shell. Figure 2c shows a cross section of the functional material layer PZT with thickness of ~1 µm. As can be seen in Figure 2d, the vibration membrane is patterned into a central circle surrounded by concentric double rings. Figure 2e depicts the parylene C deposited on the vibration membrane surface. Figure 2f shows the inclined etching gap of the vibration membrane, with a gap width of ~12 µm. Moreso, the side of a ~10-µm-thick parylene C can be clearly demonstrated, as shown in Figure 2g.

4.2. Measured SPL

Compared with the same thick rigid (SiO$_2$/Si) supporting layer, the rigid–flexible coupling one can greatly reduce structural stiffness, thereby increasing the vibration displacement and SPL.

Figure 3a shows that the SPL of the proposed piezoelectric MEMS speaker with a rigid–flexible coupling supporting layer is significantly higher than that with rigid supporting layer, and especially, is higher by 4.1–20.1 dB in the frequency range from 20 Hz to 4.2 kHz, which indicates the rigid–flexible-coupling supporting layer can improve the SPL significantly in low frequency. Moreover, a figure higher than 62 dB SPL is obtained in the frequency range of 20 Hz–20 kHz, which is at least 10 dB higher than the previous work [32], thus meeting the basic needs of human hearing [42]. Besides, an SPL of ~102.7 dB was achieved at a resonance frequency of 4 kHz. Compared with the speaker with a rigid supporting layer, the structural stiffness of the rigid–flexible coupling supporting layer prepared by us is greatly reduced, thereby making the vibration offset displacement of the vibration membrane higher, which can greatly improve the SPL at low frequency. As shown in Figure 3b, in the human audible frequency range of 20 Hz–20 kHz, a volume higher than 62 dB SPL is obtained at 2 V, and greater than 72.3 dB SPL is achieved at driving voltage of 10 V. With increasing driving voltage, the highest SPL at the resonance frequency of ~4 kHz changes from ~102.7 dB (2 V) to ~107.5 dB (4 V) to ~110.7 dB (6 V) to ~111.6 dB (8 V) and finally to ~112.5 dB (10 V).

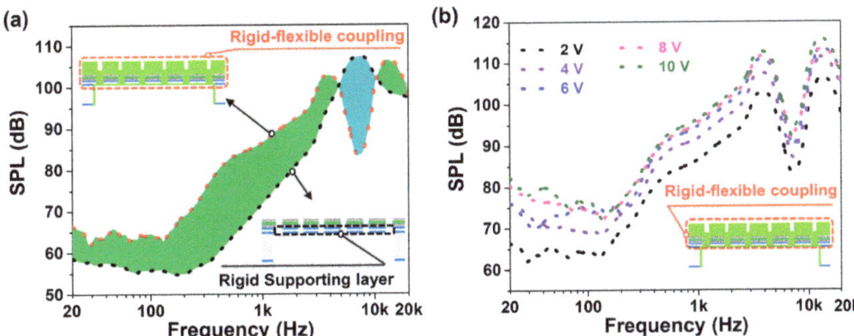

Figure 3. Measured SPL. (**a**) Comparison of SPL of a piezoelectric MEMS speaker with rigid supporting layer and rigid–flexible coupling one. (**b**) Measured SPL of piezoelectric MEMS speaker at different driving voltages.

As shown in Figure 4a–d, the SPL distribution and effective SPL at each moment when the piezoelectric MEMS speaker and a commercial electromagnetic one are used to play the same song. As can be seen, the effective SPL (L$_{Aeq}$) of the prepared piezoelectric MEMS speaker at each moment is 3–8 dB lower than that of the commercial one with the same size, and its maximum (L$_{AFmax}$) and minimum (L$_{AFmin}$) effective SPLs are also lower than those of the commercial one by 3–8 dB. However, compared with piezoelectric MEMS speaker, electromagnetic ones are difficult to miniaturize because of their larger permanent magnets [2,13,14], and also the performance of the prepared piezoelectric MEMS device meets the basic hearing needs for music, broadcast, voice interaction, etc.

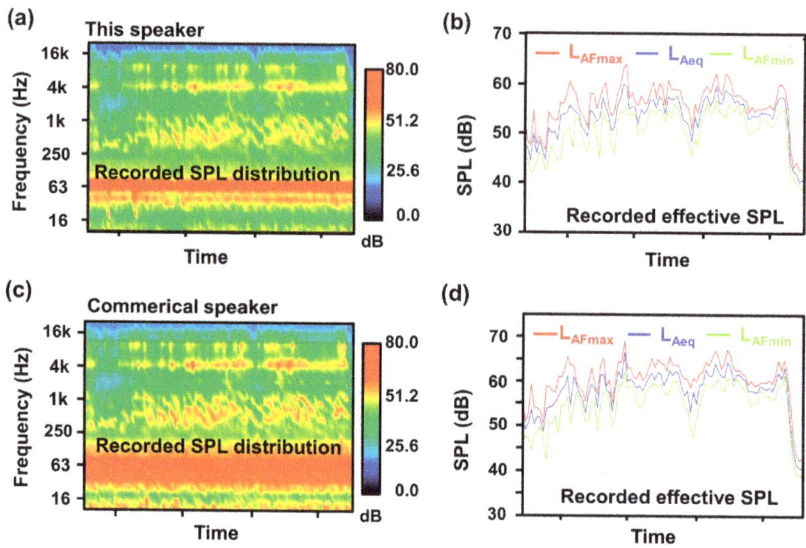

Figure 4. Comparison of a designed piezoelectric MEMS speaker and a commercial one. SPL distribution (**a,b**) effective SPL of proposed speaker at each moment when playing a song. (**c**) SPL distribution and (**d**) effective SPL of a commercial speaker at each moment when playing the same song.

4.3. Silent Alarm in Dangerous Situations

In actuality, it may not be feasible for individuals to vocalize for assistance or raise an audible alarm when confronted with a potential kidnapping or involuntary confinement, as doing so may provoke hostile parties. Thus, the development of a silent alarm sensor represents a meaningful and promising technological advancement [43,44]. In order to expand the application of piezoelectric MEMS speaker, this proposed device was attempted to be used as a silent alarm sensor, and the working mechanism is shown in Figure 5a. If we speak silently, different words produce different oral airflows that impact the membrane surface of the proposed piezoelectric MEMS device and produce different deformations, thereby generating different voltage signals. Therefore, different words can be recognized according to their unique voltage-signal characteristics. Of course, temperature also affects the voltage signals, and the increase or decrease in temperature are consistent to that of oral airflow speed, so their waveform variation law is considered to be the same. The voltage signal generated by this structure is significantly larger than the previous design work, which contributes to the accuracy of identification (Figure S2, Supplementary Materials) [36]. The distance between the mouth and the device is within ~2 cm. The corresponding valley value and number as well as overall waveform are the voltage signal characteristics being preliminary considered. Less than 2 cm is chosen as the test distance because the voltage signal generated is relatively obvious, as shown in Figure 5b. As shown in Figure 5c, for words "silent", the voltage-signal characteristics produced by the device when speaking silently with body movement are consistent with those without movement, which shows that the performance of the device is unaffected by body-movement interference. When different people speak the same word silently, the voltage-signal characteristics are consistent, which indicates the device's universal applicability to different people, as shown in Figure 5d. As can be seen in Figure 5e, when the same word is said silently, the voltage signal characteristics are also consistent in quiet and noise environments, which shows the good anti-interference performance of this device in a noisy sound environment. In Figure 5f, the words "alarm", "danger", "save" and "me" all conform to the above rules, which also well proves the universal applicability of this method. When speaking continuously, the dangerous signals are well recognized through their unique voltage signal characteristics obtained, which proves the feasibility of the device in practical application (Figure S3, Supplementary Materials). The above are only our preliminary mechanism analysis and exploration of the piezoelectric MEMS silent alarm sensor based on oral airflow, but it can also preliminarily reflect the feasibility of our proposed scheme and puts forward a new possible alternative for the traditional alarm system. Besides, this silent speech-recognition sensor can avoid the effects of external sound noise, body movement, long distance, and occlusion compared to the state of the art. Future work will combine machine learning to show remotely what they say [45,46], which is expected to be used to save people in dangerous situations, even to serve people with acquired throat injury or serious diseases, weak bodies, or who are unable to speak loudly or are in a public quiet environment. Thus, this piezoelectric MEMS speaker can also be used as a silent alarm sensor. However, we just elaborated on the working mechanism and made an attempt on its possibility, and the subsequent detailed studies are still ongoing.

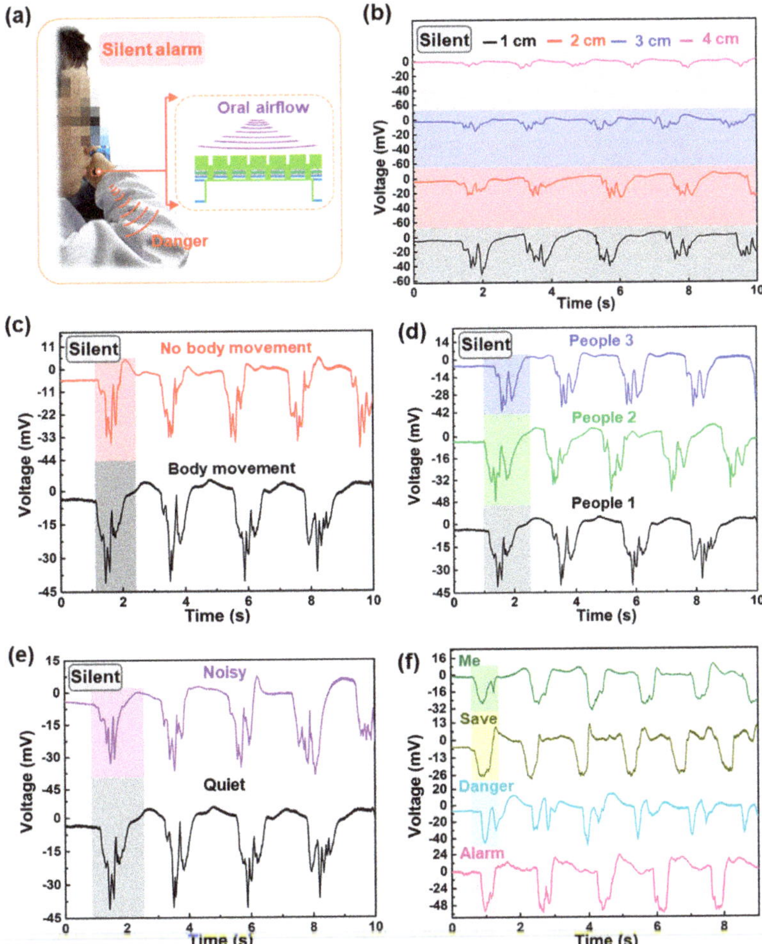

Figure 5. Silent alarm. (**a**) The wearing method and working mechanism. Unique voltage−signal characteristics of word "silent" (**b**) in different distances and (**c**) with and without body movement and (**d**) of different people when speaking the same word silently. (**e**) Anti−interference performance of device in a noisy sound environment. (**f**) Voltage-signal characteristics of the different words when speaking silently.

4.4. Compared with Previous Works

The comparisons between this work and our previous works are shown in Table 1, showing this work has its differences and advantages compared to previous works.

Table 1. Comparison between this work and our previous work.

Paper	Structure	Radius (Side Length)	Main Supporting Layer	Thickness of Parylene C	Role of Parylene C	Minimum SPL (20 Hz-20 kHz @ 2 V)	Appilications
[10]	Cantilever beam array	2 mm	Si	0.5 μm	Close the gap	~59 dB	Speaker
[33]	Improved cantilever beam array	2 mm	Si	0.5 μm	Close the gap	~55 dB	Speaker, motion monitoring, health warning
[36]	Cantilever beam array	2 mm	Si	0.2 μm	Isolate water vapor	No	Silent speech recognition
[32]	Central circle surrounded by concentric double rings	1.5 mm	Parylene C	10 μm	Supporting layer	~52 dB	Speaker
This work	Central circle surrounded by concentric double rings	2 mm	Parylene C	10 μm	Supporting layer	~62 dB	Speaker, Silent alarm

5. Conclusions

In conclusion, a piezoelectric MEMS speaker combining function of silent alarm was reviewed. The speaker mainly comprises a PZT actuation layer and a rigid–flexible coupling supporting layer. The deposition of 10 μm thick parylene C on the upper surface before etching the back cavity is the key process in forming the rigid–flexible coupling supporting layer. Additionally, the rigid vibration membrane is a central circle surrounded by concentric double rings. This proposed piezoelectric MEMS speaker, which features a rigid–flexible coupling supporting layer, can enhance the 4.1–20.1 dB SPL at low frequencies ranging from 20 Hz to 4.2 kHz, as compared to a speaker with a rigid supporting layer. In addition, this can also improve the distribution of SPL and maintain an effective SPL similar to that of a commercial electromagnetic speaker when playing the same song. The device is also capable of serving as a silent alarm in hazardous situations by detecting oral airflows. It performs exceptionally well in silently recognizing alarms based on their distinctive voltage-signal features, while effectively avoiding the impacts of external noise, body movement, long distance, and blockage. Thus, both the piezoelectric and inverse piezoelectric effects of PZT thin film can be utilized. This prepared device provides a reference for the functional expansion of piezoelectric MEMS speakers.

Supplementary Materials: The following supporting information can be downloaded at: https://www.mdpi.com/article/10.3390/mi14030702/s1.

Author Contributions: Conceptualization, Q.W. and T.R.; methodology, Q.W.; software, Q.W.; validation, Q.W., T.R. and Q.X.; formal analysis, Q.W. and Z.H.; investigation, Q.W.; resources, Q.W.; data curation, Q.W.; writing—original draft preparation, Q.W.; writing—review and editing, B.Y., Z.L. and M.Y.; visualization, Q.W.; supervision, Z.L. and J.L.; project administration, J.L.; funding acquisition, J.L. All authors have read and agreed to the published version of the manuscript.

Funding: This work was partially supported by the National Natural Science Foundation of China (No. 42127807-03), the National Key R&D Program of China under grant (2020YFB1313502), the Strategic Priority Research Program of Chinese Academy of Sciences (Grant No. XDA25040100, XDA25040200 and XDA25040300), Shanghai Pilot Program for Basic Research—Shanghai Jiao Tong University (No. 21TQ1400203), SJTU Trans-med Award (No. 2019015, 21X010301627), the Oceanic Interdisciplinary Program of Shanghai Jiao Tong University (No. SL2020ZD205, SL2020MS017, SL2103), and the Scientific Research Fund of Second Institute of Oceanography, MNR (No. SL2020ZD205).

Data Availability Statement: Not applicable.

Acknowledgments: The authors are also grateful to the Center for Advanced Electronic Materials and Devices (AEMD) of Shanghai Jiao Tong University.

Conflicts of Interest: The authors declare no conflict of interest.

References

1. Wang, H.R.; Ma, Y.F.; Zheng, Q.C.; Cao, K.; Lu, Y.; Xie, H.K. Review of Recent Development of MEMS Speakers. *Micromachines* **2021**, *12*, 1257. [CrossRef]
2. Wang, H.R.; Chen, Z.F.; Xie, H.K. A High-SPL Piezoelectric MEMS Loud Speaker Based on Thin Ceramic PZT. *Sens. Actuator A Phys.* **2020**, *309*, 112018. [CrossRef]
3. Je, S.-S.; Kim, J.; Harrison, J.C.; Kozicki, M.N.; Chae, J. In Situ Tuning of Omnidirectional Microelectromechanical-Systems Microphones to Improve Performance Fit in Hearing Aids. *J. Appl. Phys. Lett.* **2008**, *93*, 123501. [CrossRef]
4. Woo, S.T.; Han, J.-H.; Lee, J.H.; Cho, S.; Seong, K.-W.; Choi, M.; Cho, J.-H. Realization of a High Sensitivity Microphone for a Hearing Aid Using a Graphene–PMMA Laminated Diaphragm. *ACS Appl. Mater. Interfaces* **2017**, *9*, 1237–1246. [CrossRef] [PubMed]
5. Babu, A.; Malik, P.; Das, N.; Mandal, D. Surface Potential Tuned Single Active Material Comprised Triboelectric Nanogen-erator for a High Performance Voice Recognition Sensor. *Small* **2022**, *18*, 2201331. [CrossRef]
6. Aloysius, N.; Geetha, M. Understanding Vision-Based Continuous Sign Language Recognition. *Multimed. Tools Appl.* **2020**, *79*, 22177–22209. [CrossRef]
7. Ji, Y.; Liu, L.; Wang, H. Updating the Silent Speech Challenge Benchmark with Deep Learning. *Speech Commun.* **2018**, *98*, 42–50. [CrossRef]
8. Wang, Y.; Chao, M.; Wan, P.; Zhang, L. A Wearable Breathable Pressure Sensor From Metal-Organic Framework Derived Nanocomposites for Highly Sensitive Broad-Range Healthcare Monitoring. *Nano Energy* **2020**, *70*, 104560. [CrossRef]
9. Yu, Q.Y.; Zhang, P.; Chen, Y.C. Human Motion State Recognition Based on Flexible, Wearable Capacitive Pressure Sensors. *Micromachines* **2021**, *12*, 1219. [CrossRef]
10. Wang, Q.; Ruan, T.; Xu, Q.D.; Shi, Y.Z.; Yang, B.; Liu, J.Q. Wearable Multifunctional Piezoelectric MEMS Device for Motion Monitoring, Health Warning, and Earphone. *Nano Energy.* **2021**, *89*, 106324. [CrossRef]
11. Khan, Y.; Ostfeld, A.E.; Lochner, C.M.; Pierre, A.; Arias, A.C. Monitoring of Vital Signs with Flexible and Wearable Medical Devices. *Adv. Mater.* **2016**, *28*, 4373–4395. [CrossRef] [PubMed]
12. Qiu, Y.; Sun, S.; Xu, C.; Wang, Y.; Tian, Y.; Liu, A.; Hou, X.; Chai, H.; Zhang, Z.; Wu, H. The Frequency-Response Behaviour of Flexible Piezoelectric Devices for Detecting the Magnitude and Loading Rate of Stimuli. *J. Mater. Chem. C* **2021**, *9*, 584–594. [CrossRef]
13. Rehder, J.; Rombach, P.; Hansen, O. Magnetic Flux Generator for Balanced Membrane Loudspeaker. *Sens. Actuator A Phys.* **2002**, *97*, 61–67. [CrossRef]
14. Shahosseini, I.; Lefeuvre, E.; Moulin, J.; Martincic, E.; Woytasik, M.; Lemarquand, G. Optimization and Microfabrication of High Performance Silicon-Based MEMS Microspeaker. *IEEE Sens. J.* **2013**, *13*, 273–284. [CrossRef]
15. Garud, M.V.; Pratap, R. A Novel MEMS Speaker with Peripheral Electrostatic Actuation. *J. Microelectromech. Syst.* **2020**, *29*, 592–599. [CrossRef]
16. Langa, B.S.; Ehrig, L.; Stolz, M.; Schuffenhauer, D. Concept and Proof for an All-Silicon MEMS Micro Speaker Utilizing Air Chambers. *Microsyst. Nanoeng.* **2019**, *5*, 43.
17. Ozdogan, M.; Towfighian, S.; Miles, R.N. Modeling and Characterization of a Pull-in Free MEMS Microphone. *IEEE Sens. J.* **2020**, *20*, 6314–6323. [CrossRef]

18. Ali, W.R.; Prasad, M. Piezoelectric MEMS Based Acoustic Sensors: A Review. *Sens. Actuator A-Phys.* **2020**, *301*, 31. (In English) [CrossRef]

19. Kim, S.; Zhang, X.; Daugherty, R.; Lee, E.; Kunnen, G.; Allee, D.R.; Forsythe, E. Chae, Microelectromechanical Systems (MEMS) Based-Ultrasonic Electrostatic Actuators on a Flexible Substrate. *IEEE Electron Device Lett.* **2012**, *33*, 1072–1074. [CrossRef]

20. Wang, C.; Hosomi, T.; Nagashima, K.; Takahashi, T.; Zhang, G.; Kanai, M.; Yoshida, H.; Yanagida, T. Phosphonic Acid Modified ZnO Nanowire Sensors: Directing Reaction Pathway of Volatile Carbonyl Compounds. *ACS Appl. Mater. Interfaces.* **2020**, *12*, 44265–44272. [CrossRef]

21. Fei, C.; Liu, X.; Zhu, B.; Li, D.; Yang, X.; Yang, Y.; Zhou, Q. AlN Piezoelectric Thin Films for Energy Harvesting and Acoustic Devices. *Nano Energy* **2018**, *51*, 146–161. [CrossRef]

22. George, J.P.; Smet, P.F.; Botterman, J.; Bliznuk, V.; Woestenborghs, W.; Van Thourhout, D.; Neyts, K.; Beeckman, J. Lanthanide-Assisted Deposition of Strongly Electro-optic PZT Thin Films on Silicon: Toward Integrated Active Nanophotonic Devices. *ACS Appl. Mater. Interfaces.* **2015**, *7*, 13350–13359. [CrossRef] [PubMed]

23. Park, K.I.; Son, J.H.; Hwang, G.T.; Jeong, C.K.; Ryu, J.; Koo, M.; Choi, I.; Lee, S.H.; Byun, M.; Wang, Z.L.; et al. Highly-Efficient, Flexible Piezoelectric PZT Thin Film Nanogenerator on Plastic Substrates. *Adv. Mater.* **2014**, *26*, 2514–2520. [CrossRef]

24. Hwang, G.T.; Park, H.; Lee, J.H.; Oh, S.; Park, K.I.; Byun, M.; Park, H.; Ahn, G.; Jeong, C.K.; No, K.; et al. Self-Powered Cardiac Pacemaker Enabled by Flexible Single Crystalline PMN-PT Piezoelectric Energy Harvester. *Adv. Mater.* **2014**, *26*, 4880–4887. [CrossRef] [PubMed]

25. Xu, S.; Yeh, Y.-w.; Poirier, G.; McAlpine, M.C.; Register, R.A.; Yao, N. Flexible Piezoelectric PMN–PT Nanowire-Based Nanocomposite and Device. *Nano Lett.* **2013**, *13*, 2393–2398. [CrossRef]

26. Wu, F.; Cai, W.; Yeh, Y.-W.; Xu, S.; Yao, N. Energy Scavenging Based on A Single-Crystal PMN-PT Nanobelt. *Sci. Rep.* **2016**, *6*, 22513. [CrossRef]

27. Yang, Z.; Zu, J. Comparison of PZN-PT, PMN-PT Single Crystals and PZT Ceramic for Vibration Energy Harvesting. *Energy Convers. Manag.* **2016**, *122*, 321–329. [CrossRef]

28. Gao, X.; Yang, J.; Wu, J.; Xin, X.; Li, Z.; Xiaoting, Y.; Shen, X.; Dong, S. Piezoelectric Actuators and Motors: Materials, Designs, and Applications. *Adv. Mater. Technol.* **2019**, *5*, 1900716. [CrossRef]

29. Cheng, H.; Lo, S.; Wang, Y.; Chen, Y.; Lai, W.; Hsieh, M.; Wu, M.; Fang, W. Piezoelectric Microspeaker Using Novel Driving Approach and Electrode Design for Frequency Range Improvement. In Proceedings of the IEEE International Conference on Micro Electro Mechan-ical Systems (MEMS), Vancouver, BC, Canada, 18–22 January 2020; pp. 513–516.

30. Nguyen, M.D.; Houwman, E.P.; Rijnders, G. Large Piezoelectric Strain with Ultra-Low Strain Hysteresis in Highly C-Axis Oriented Pb(Zr0.52Ti0.48)O3 Films with Columnar Growth on Amorphous Glass Substrates. *Sci. Rep.* **2017**, *7*, 12915. [CrossRef]

31. Xu, Q.; Zhang, G.; Ding, J.; Wang, R.; Pei, Y.; Ren, Z.; Shang, Z.; Xue, C.; Zhang, W. Design and Implementation of Two-Component Cilia Cylinder MEMS Vector Hydrophone. *Sens. Actuator A Phys.* **2018**, *277*, 142–149. [CrossRef]

32. Wang, Q.; Ruan, T.; Xu, Q.D.; Shi, Y.Z.; Yang, B.; Liu, J.Q. Piezoelectric MEMS Speaker with Rigid-Flexible-Coupling Actuation Layer. In Proceedings of the IEEE International Conference on Micro Electro Mechanical Systems (MEMS), Tokyo, Japan, 9–13 January 2022; pp. 620–623.

33. Wang, Q.; Yi, Z.; Ruan, T.; Xu, Q.; Yang, B.; Liu, J. Obtaining High SPL Piezoelectric MEMS Speaker via a Rigid-Flexible Vibration Coupling Mechanism. *J. Microelectromech. Syst.* **2021**, *30*, 725–732. [CrossRef]

34. Barker, J.P.; Marxer, R.; Vincent, E. Multi-Microphone Speech Recognition in Everyday Environments. *Comput. Speech Lang.* **2017**, *46*, 386–387. [CrossRef]

35. Zhu, M.; Huang, Z.; Wang, X. Automatic Speech Recognition in Different Languages Using High-Density Surface Electromyography Sensors. *IEEE Sens. J.* **2020**, *99*, 14155–14167.

36. Wang, Q.; Ruan, T.; Xu, Q.D.; Shi, Y.Z.; Yang, B.; Liu, J.Q. Piezoelectric Mems Unvoiced Speech-Recognition Sensor Based on Oral Airflow. In Proceedings of the IEEE International Conference on Micro Electro Mechanical Systems (MEMS), Tokyo, Japan, 9–13 January 2022; pp. 63–66.

37. Wang, H.; Godara, M.; Chen, Z.; Xie, H. A One-Step Residue-Free Wet Etching Process of Ceramic PZT For Piezoelectric Transducers. *Sens. Actuator A Phys.* **2019**, *290*, 130–136. [CrossRef] [PubMed]

38. Zhao, H.J.; Liu, J.-S.; Ren, T.; Liu, Y.X.; Liu, L.T.; Li, Z.J. Study of Fabrication and Etching Processes of PZT Thin Films on Silicon. *Piezoelectrics Acoustooptics* **2001**, *23*, 290–292.

39. Kuo, W.-C.; Chen, C.-W. Fabrication of Parylene-Based High-Aspect-Ratio Suspended Structure Using a Silicon-on-Insulator Wafer. *Jpn. J. Appl. Phys.* **2013**, *52*, 036501. [CrossRef]

40. Ortigoza-Diaz, J.; Scholten, K.; Larson, C.; Cobo, A.; Hudson, T.; Yoo, J.; Baldwin, A.; Weltman Hirschberg, A.; Meng, E. Techniques and Considerations in the Microfabrication of Parylene C Microelectromechanical Systems. *Micromachines* **2018**, *9*, 422. [CrossRef] [PubMed]

41. Kim, B.J.; Meng, E. Micromachining of Parylene C for BioMEMS. *Polymer. Adv. Technol.* **2016**, *27*, 564–576. [CrossRef]

42. Jin, S.W.; Jeong, Y.R.; Park, H.; Keum, K.; Lee, G.; Lee, Y.H.H.; Kim, M.S.; Ha, J.S. A Flexible Loudspeaker Using the Movement of Liquid Metal Induced by Electrochemically Controlled Interfacial Tension. *Small* **2019**, *15*, 1905263. [CrossRef]

43. Gonzalez-Lopez, J.A.; Gomez-Alanis, A.; Martín-Doas, J.M.; Pérez-Córdoba, J.L.; Gomez, A.M. Silent Speech Interfaces for Speech Restoration: A Review. *IEEE Access* **2020**, *8*, 177995. [CrossRef]

44. Yoo, H.; Kim, E.; Chung, J.W.; Cho, H.Y.; Jeong, S.; Kim, H.; Jang, D.; Kim, H.; Yoon, J.; Lee, G.H.; et al. Silent Speech Recognition with Strain Sensors and Deep Learning Analysis of Directional Facial Muscle Movement. *ACS Appl. Mater. Interfaces* **2022**, *14*, 54157–54169. [CrossRef] [PubMed]
45. Sahs, J.; Khan, L. A Machine Learning Approach to Android Malware Detection. In Proceedings of the Intelligence and Security Informatics Conference (EISIC), Odense, Denmark, 22–24 August 2012; pp. 141–147.
46. Xiao, W.; Wang, D.; Bao, L. Emotional State Classification from EEG Data Using Machine Learning Approach. *Neurocomputing* **2014**, *129*, 94–106.

MDPI

Article

Cluster-ID-Based Throughput Improvement in Cognitive Radio Networks for 5G and Beyond-5G IoT Applications

Stalin Allwin Devaraj [1], **Kambatty Bojan Gurumoorthy** [2], **Pradeep Kumar** [3], **Wilson Stalin Jacob** [4], **Prince Jenifer Darling Rosita** [5] and **Tanweer Ali** [6,*]

[1] Department of Electronics and Communication Engineering, Francis Xavier Engineering College, Tirunelveli 627003, India
[2] Department of Electronics and Communication Engineering, KPR Institute of Engineering and Technology, Coimbatore 641407, India
[3] Discipline of Electrical, Electronic and Computer Engineering, University of KwaZulu-Natal, Durban 4041, South Africa
[4] Engineering Department—Electrical Engineering, Botho University, Gaborone 501564, Botswana
[5] Electrical Engineering Department, New Era College, Gaborone 501564, Botswana
[6] Department of Electronics and Communication Engineering, Manipal Institute of Technology, Manipal Academy of Higher Education, Manipal 576104, India
* Correspondence: tanweer.ali@manipal.edu

Abstract: Cognitive radio (CR), which is a common form of wireless communication, consists of a transceiver that is intelligently capable of detecting which communication channels are available to use and which are not. After this detection process, the transceiver avoids the occupied channels while simultaneously moving into the empty ones. Hence, spectrum shortage and underutilization are key problems that the CR can be proposed to address. In order to obtain a good idea of the spectrum usage in the area where the CRs are located, cooperative spectrum sensing (CSS) can be used. Hence, the primary objective of this research work is to increase the realizable throughput via the cluster-based cooperative spectrum sensing (CBCSS) algorithm. The proposed scheme is anticipated to acquire advanced achievable throughput for 5G and beyond-5G Internet of Things (IoT) applications. Performance parameters, such as achievable throughput, the average number of clusters and energy, have been analyzed for the proposed CBCSS and compared with optimal algorithms.

Keywords: cluster-based cooperative spectrum sensing; achievable throughput; greedy heuristic algorithm; cognitive radio network; 5G and beyond-5G IoT applications

Citation: Devaraj, S.A.; Gurumoorthy, K.B.; Kumar, P.; Jacob, W.S.; Rosita, P.J.D.; Ali, T. Cluster-ID-Based Throughput Improvement in Cognitive Radio Networks for 5G and Beyond-5G IoT Applications. *Micromachines* **2022**, *13*, 1414. https://doi.org/10.3390/mi13091414

Academic Editors: Qiongfeng Shi and Huicong Liu

Received: 2 July 2022
Accepted: 25 August 2022
Published: 28 August 2022

Publisher's Note: MDPI stays neutral with regard to jurisdictional claims in published maps and institutional affiliations.

1. Introduction

Cognitive radio (CR) is a relatively new long-haul radio innovation. After the software-defined radio (SDR), which is, to a greater degree, gradually becoming a reality, CR will be the more successful radio correspondence framework to be created [1–10]. The principal focal point of this research work is the range designation issue, i.e., to accurately choose the Secondary Users (SUs) to detect and use the Primary User (PU) channels for a wide variety of situations [11–22].

Since the CR is being used to provide a strategy for utilizing the range more effectively, range detection is a significant issue in 5G and beyond-5G IoT applications, and the frequency band used is narrow band LTE (NB-LTE) [23–25]. The general framework is to work adequately and to provide the necessary enhancement in range detection. The CR range detection framework must have the option to successfully distinguish the additional transmissions, recognize them and enlighten the central processing unit of the CR such that the crucial move can be made.

This work addresses the following:

- Methods of increasing the throughput of SUs.

- A discussion of the spectrum utilization problem.
- The proposal of cluster-based cooperative spectrum sensing (CBCSS) to perform cooperative spectrum sensing in 5G and beyond-5G IoT applications.

To extract the minimum-sensing users, a closed-form expression is logically impractical, and a number of security issues need to be considered. A higher dimensions vector is not considered here since it increases the computational cost and makes the algorithm undesirable.

2. Literature Survey

The extraction of the number of minimum-sensing users in a closed-form expression is analytically unfeasible. Therefore, the analytical formulation of the amount of saved energy in each scenario must be considered. In addition, a greater number of security issues are considered. The higher dimensions vector is not considered, and if considered, it increases the computational cost, which makes the algorithm undesirable. There are also practical protocol issues of synchronization and the estimation and tracking of the traffic parameters. In addition to spectrum sensing to improve spectrum utilization, a CR in a cognitive radio network (CRN) can sense available networks and communication systems around it. The CRNs are composed of various kinds of communication systems and networks and can be viewed as a sort of heterogeneous network. Dusit Niyato et al. recommended a Bertrand game model to analyze the impacts of system parameters such as spectrum substitutability and channel [7]. Nasif et al. offered an algorithm for opportunistic spectrum sharing with multiple co-channel primary transmitters. The authors presented a distributed collaborative algorithm for cognitive radios [14]. Quan et al. proposed a soft computing-based algorithm that requires full consideration of the noise level of all secondary users. The throughput performance of SUs in cognitive radio networks has been analyzed [17]. Yue Wang et al. proposed an anti-jamming problem in the presence of a smart jammer. This smart jammer learns the transmission power of the user and adjusts its transmission power to maximize the damaging effect that is being analyzed [4,21]. Mohammad Rashid et al. developed a framework to learn the QoS performance measures using a queuing analysis in the data link layer for infrastructure-based cognitive radio users in the opportunistic spectrum access [12]. Song et al. considered the interference temperature restriction and opportunistic spectrum allocation and suggested a suitable framework for the joint spectrum allocation and power control to make use of the utilized and underutilized licensed electromagnetic spectrum [19].

Bin Wang et al. proposed an approach to enhance the performance of unlicensed users by utilizing the licensed user spectrum in cognitive radio networks [3]. Minh-Viet Nguyen et al. investigate the problem of resource allocation spectrum sensing methods, frequency selectivity, spectrum allocation and frequency bands in cognitive radio relay networks [11]. Azarfar et al. proposed an auction approach along with an anticipated double spectrum to augment the spectrum utilization in a CR network [2]. Li et al. recommended a new cooperative spectrum sensing framework that would effectively combine spectrum sensing and spectrum sharing [10]. Ruby et al. introduced a new hard two-bit overhead for each individual user. Consequently, a balanced trade-off between the detection performance and complexity was realized [18]. Dongyue Xue et al. proposed a scheduling algorithm to reduce the spectrum sensing time for wireless mesh applications [6]. Changpeng Ji et al. proposed a cross-layer cluster-based spectrum sensing algorithm to achieve better noise density and slot length [5]. Li Yu et al. proposed a framework in order to increase the performance of a CR network by using the cluster-based cooperative spectrum sensing technique [9]. Parzy et al. employed a distributed resource allocation mechanism for CR networks based on a novel competition methodology, which syndicates the benefits of node competition and cooperation [16]. Nadine Abbas et al. presented the spectrum availability and scarcity in the radio spectrum for cognitive radio networks [13]. Tsakmalis et al. extended an algorithm to increase the cognitive radio network throughput and reduce the primary user interference through scheduling techniques [20]. Zhu et al. considered the resource allocation issue in an OFDMA-based cognitive radio network and increased the coverage area of the antenna using several beam forming techniques to support the secondary user in using the licensed user spectrum [22]. Orumwense et al. proposed a cooperative technique to address the spectrum sensing issues of secondary users. The primary user channel length, spectrum sensing time and slot length were all measured for the spectrum allocation decision for secondary users [15]. Alberti et al. concluded that the problem of sensor applications in cognitive radio networks is that the parameters considered provide a short network lifetime and poor beamforming characteristics [1]. Heejung et al. presented a survey on the next generation of Internet of Things (IoT) networks, showing an increase in delay, network lifetime, packet delivery ratio and throughput [8].

In this paper, the heterogeneities in both PU channels and SUs are investigated. The PU channel is characterized by channel idle probability and channel capacity, while the SU is depicted by the energy detection threshold, received SNR and geographical location.

3. Cluster-Based Cooperative Spectrum Sensing (CBCSS)

System Model

Consider a CRN with N SUs and M PU channels. Each channel is exclusively used by the PU. However, the PU is idle, and the SU can opportunistically utilize the channel when it is available through spectrum sensing. Let M be the set of such PU channels and N denote the set of SUs. Figure 1 demonstrates that the channel heterogeneity-spectrum availability varies across the SUs. The SUs that are located far from the PU will only report noise when the detection range of the PUs only covers part of the system. Hence, the CRN is partitioned into clusters so that the SUs in each cluster are within the detection range of the same set of PU channels.

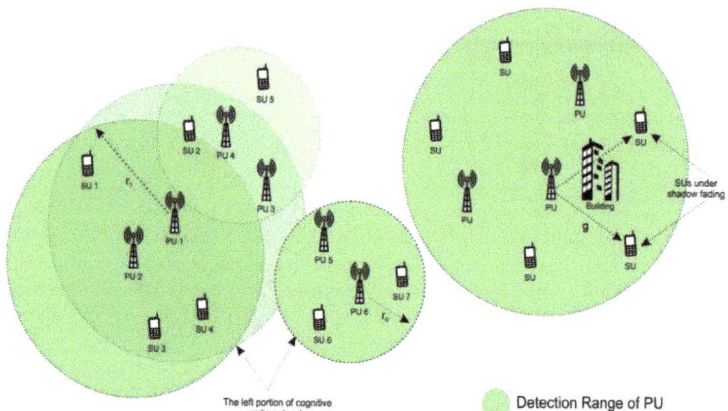

Figure 1. System model for CBCSS.

The vacant portion of the spectrum can only be allotted by a CR user. We use a sampling frequency f_s to sample the frequency.

$$H_j^1 : y_{i,j}(k) = s_{i,j}(k) + u_{i,j}(k) \; i = 1, 2, \ldots \ldots, N \tag{1}$$

$$H_j^0 : y_{i,j}(k) = u_{i,j}(k) \; j = 1, 2, \ldots \ldots, N \tag{2}$$

The false alarm probability $P_{f,(i,j)}$ is defined as the probability of the SU j under H_j^0, which is given by

$$P_{f,(i,j)} = Q\left(\left(\frac{\epsilon_i}{\sigma_{u_{i,j}}^2} - 1\right)\sqrt{f_s\,T}\right) \tag{3}$$

The detection probability $P_{d,(i,j)}$ is defined as

$$P_{d,(i,j)} = Q\left(\left(\frac{\epsilon_i}{\sigma_{u_{i,j}}^2} - 1 - \gamma_{i,j}\right)\sqrt{\frac{f_s T}{2\gamma_{i,j} + 1}}\right) \tag{4}$$

In order to provide sufficient protection to the Pus, it is required to keep the detection probability above a given threshold Q_{th}, that is, $Q_{d,j} \geq Q_{th}$. Hence,

$$\prod_{i=1}^{m}(1 - P_{d,(i,j)}) \geq Q_{th} \tag{5}$$

The SUs and PU channel's allocation matrices are $[X_s]_{N \times K}$ and $[X_c]_{M \times K}$. The elements $x_{s,i}^k$ and $x_{c,j}^k$ can be defined as:

$$x_{s,i}^k \begin{cases} 1 \; if \; SU \; is \; with \; cluster \\ \qquad r \\ 0 \; otherwise \end{cases} \tag{6}$$

$$x_{c,j}^k \begin{cases} 1 \; if \; CH_j \; is \; with \; cluster \\ \qquad r \\ 0 \; otherwise \end{cases} \tag{7}$$

Consider the following two vectors:
S_k represents the set of SUs in cluster k

$$S_k = \{i \mid x_{s,i}^k = 1, \forall i \in N\} \tag{8}$$

B_k denotes the set of PU channels sensed and utilized by the SUs in cluster k

$$B_k = \{j \mid x_{c,j}^k = 1, \forall j \in M\} \tag{9}$$

Thus, the total throughput is given by

$$R_k(S_k, B_k) = \sum_{j \in B_k} \frac{T - \tau}{T} P(H_j) C_j (1 - Q_{f,j}^k(S_k, B_k)) \tag{10}$$

where $P(H_j)$ is the idle probability for channel j, C_j is the transmission capacity for channel j, and

$$Q_{f,j}^k(S_k, B_k) = 1 - \prod_{i \in S_k}(1 - P_{f,(i,j)}(\frac{\tau}{b_k})) \tag{11}$$

To represent the assignment policy, a three-dimensional matrix A_{NXMXK} is defined as

$$A_{ijk}^n \begin{cases} 1 \; if \; i \in S_k \; and \; j \in B_k \\ \qquad - \\ 0 \; otherwise \end{cases} \tag{12}$$

The problem is formulated and given as

$$\max_{X_s X_c,} = \sum_k R_k(S_k(X_s), B_k(X_c)) \tag{13}$$

$$\sum_{k=1}^{K} x_{s,i}^k = 1, \forall i \tag{14}$$

$$\sum_{k=1}^{K} x_{c,j}^k = 1, \forall j \tag{15}$$

$$\sum_{i \in S_k} x_{s,i}^k \geq \overline{m}, \forall k \tag{16}$$

The optimization problem can be solved in polynomial time if and only if the corresponding decision problem can be solved. Thus, the proof of the algorithm for an optimization problem is equivalent to proving its corresponding decision problem.

4. Results and Discussion

Simulation Parameters

The simulation results have been presented for the proposed cluster-based cooperative spectrum sensing with a greedy heuristic Algorithm 1 (GHA) based on the analytical expressions established in the previous section. The performance features such as the achievable throughput, average count of clusters and energy of the proposed cluster-based greedy heuristic algorithm have been appraised and linked with the conventional optimal algorithm using MATLAB. The assumptions made in the study are given in Table 1.

Algorithm 1: Greedy algorithm

Input:

$G_K (X_k \cup Y_k, \in_k),. m_k$

Initialization: $S_k = X_k$, $B_k = \varnothing$, $A^{(k-1)}, 1 \leftarrow 1$, $X_s^k = [1]$ and $X_c^k = [0]$

$y_1 = \arg \max_{y \in Y_k} \deg(y)$

while $|S_k| \geq m_k$ and $|Y_k| > 0$

$y_1 = \arg \max_{y \in \frac{Y_k}{B_k}}$

ifdeg$(y)m_k$**then**

break;

else

$P_k = 1$;

$B_k \leftarrow B_k \cup y_1$; $\Gamma_k[1] = R_k(S_k, B_k)$;

$x_{c,y_1}^k = 1$;

$x_{s,i}^k = 0$; $\forall i \in \psi_{y_{1r}}$; update $A_{i\,j\,k}$;

end if

Find $1^* = \arg \max_1 \Gamma_k[1]$;

$S_k = \cap_{1=1}^{1^*} \psi_{y_{1r}}; B_k = \left\{ CH_{P_k[1]}, CH_{P_k[2]} \cdots \cdots \cdots \cdots \cdots \cdots CH_{P_k[1*]} \right\}.$

return rules

Table 1. Simulation parameters.

Description	Range
Simulation Area	1200 × 700 m
Primary Users (M)	50
Secondary Users (N)	40
Cluster Size	25 users/cluster
Number of Clusters	04
Transmission Range	250 m
Packet Size	165 bytes
Mobility Model	Random Waypoint
Node Pause Time	5 s
Sampling Frequency	6 MHz
Sensing Time	3–4 ms

In Figure 2, the node setup vs. cluster formation for the proposed CBCSS is plotted. The simulation area of 1200 m is used for the X-axis, and 700 m is used for the Y-axis for the proposed CBCSS.

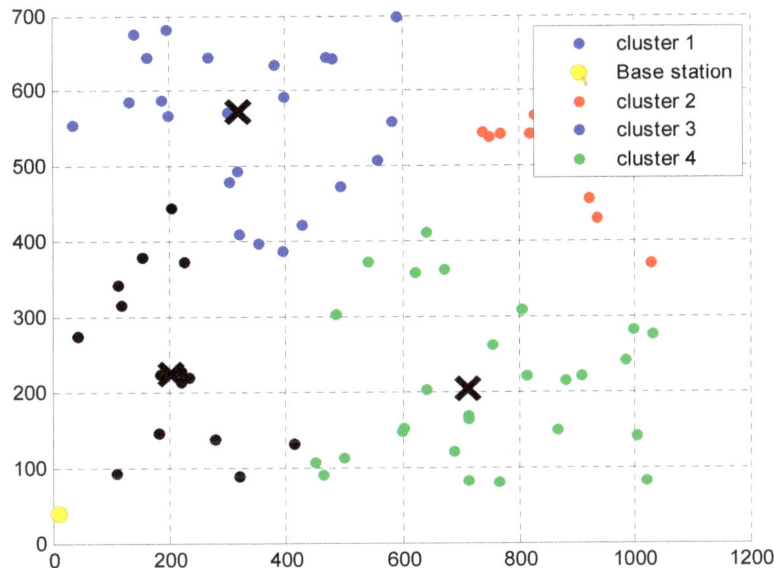

Figure 2. Simulation node setup vs. cluster formation.

Figure 3 shows the total sensed PU channels (M) vs. the achievable throughput graphs for diverse values N = 20, N = 25, N = 30, where N represents the number of secondary channels. The number of sensed PU channels (M) is used for the X-axis, and the attainable throughput is used for the Y-axis.

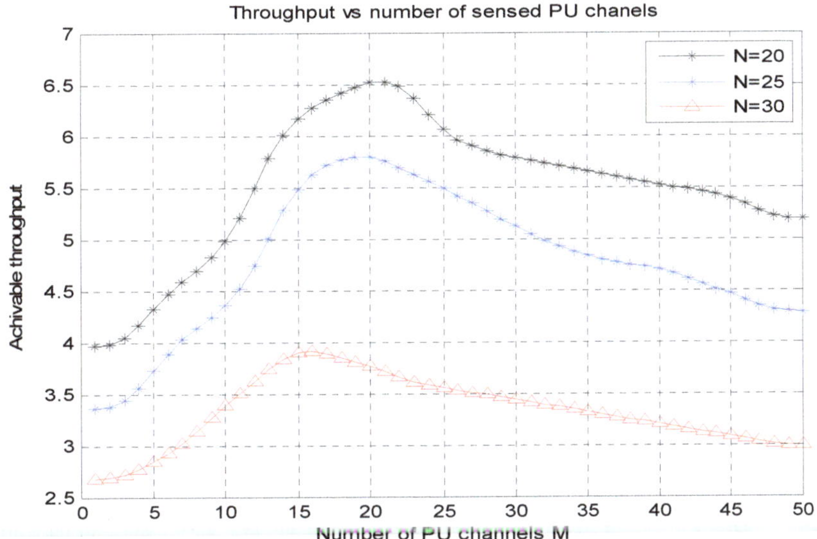

Figure 3. Number of sensed PU channels (M) vs. achievable throughput.

Figure 4 shows the amount of PU channels (M) vs. the average figure of cluster graphs for N = 20, N = 25, N = 30. The number of PU channels (M) is used for the X-axis, and the average number of clusters is used for the Y-axis.

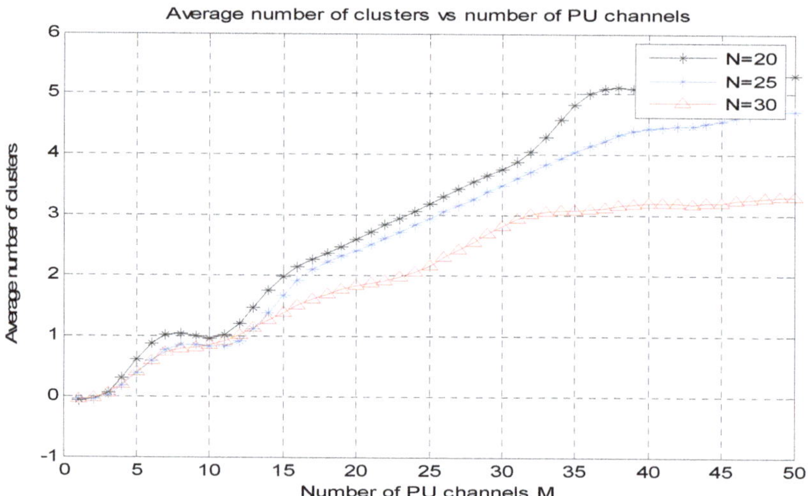

Figure 4. Number of PU channels (M) vs. average number of clusters.

It is apparent that when the number of PU channels rises, the average figure of clusters also rises because the number of clusters formed relies heavily on the detection range of the PU.

Figure 5 shows the count of the PU channel (M) vs. achievable throughput graphs for optimal N = 20, greedy N = 20, optimal N = 40 and greedy N = 30 ranges. The number of PU channels (M) is used for the X-axis, and the achievable throughput is used correspondingly for the Y-axis.

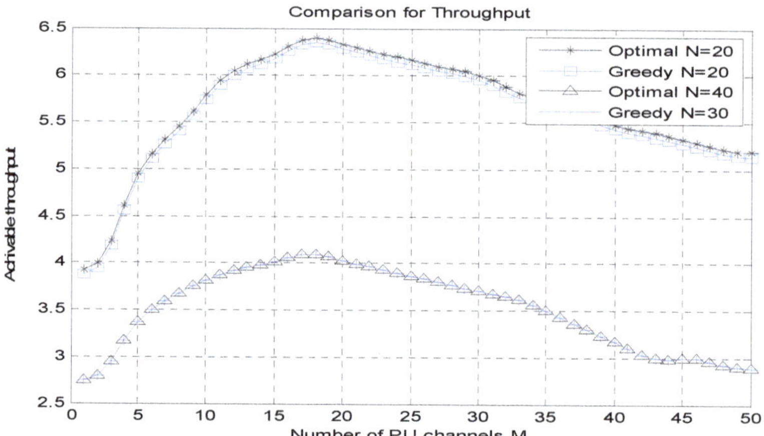

Figure 5. Number of PU channels (M) vs. achievable throughput.

It has been perceived that the suggested algorithm attains an almost ideal performance, with an extreme performance loss of 4.6% for the achievable throughput.

Figure 6 shows the number of iterations vs. energy (bits/J) graph for optimal and greedy algorithm techniques. The number of iterations is used for the X-axis, and the energy is used correspondingly for the Y-axis.

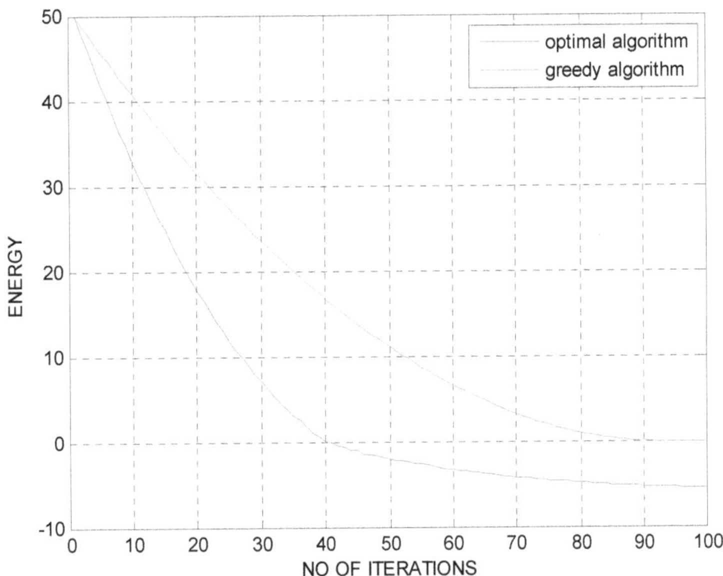

Figure 6. Number of iterations vs. energy (bits/J).

From the comparison graphs, it has been detected that the energy saved is expressed as a function of the number of iterations. Clearly, the number of iterations surges as the energy declines. The time necessary for sensing the PU channel is known as an iteration.

The results of the comparison between the proposed system and the existing systems are shown in Table 2. The simulation parameters, such as achievable throughput, the average number of clusters and energy, have been associated.

Table 2. The comparison of the existing methods and proposed CBCSS.

Parameters	Song et al. [7]	Ruby et al. [12]	Li et al. [15]	Nadine et al. [17]	Proposed CBCSS
Achievable throughput	92.73%	84.83%	90.24%	82.90%	96.87%
Average number of clusters	2	1	3	3	5
Energy	1.87 bits/J	3.64 bits/J	0.94 bits/J	6.23 bits/J	0.23 bits/J

Table 2 reveals that the proposed CBCSS for 5G and beyond 5G IoT applications has high achievable throughput (96.87%), a high average number of clusters (5) and less energy (0.23 bits/J) than the existing schemes. Based on the simulation result, it is concluded that the proposed CBCSS algorithm provides a better solution for the SUs and is suitable for utilizing the PU channels effectively in 5G and beyond-5G IoT applications. Thus, the overall performance analysis suggests that the CBCSS algorithm for the cooperative spectrum sensing with a greedy heuristic algorithm for 5G and beyond-5G IoT applications in CRN performed well and achieved good performance in terms of achievable throughput, the average number of clusters and energy. Hence, for the effective 5G and beyond-5G IoT communication applications, the CBCSS scheme can enable additional benefits, such as the maximum achievable throughput.

5. Conclusions

In order to capitalize on the achievable throughput of cognitive radio networks, a CBCSS with GHA has been offered for 5G and beyond-5G IoT applications. The CBCSS algorithm was suitably developed. Performance parameters, such as achievable throughput, the average number of clusters and energy, have been scrutinized for the proposed CBCSS algorithm. In comparison with the ideal algorithm, the recommended CBCSS with GHA showed an achievable throughput of 96.87%. Thus, the proposed greedy algorithm performed better in terms of high achievable throughput and low energy.

Author Contributions: Conceptualization, S.A.D., K.B.G., P.K., W.S.J., P.J.D.R. and T.A.; methodology, S.A.D., K.B.G., W.S.J., P.J.D.R.; software, S.A.D., K.B.G., W.S.J.; validation, S.A.D., K.B.G. and W.S.J.; formal analysis, S.A.D.; investigation, S.A.D. and K.B.G.; resources, S.A.D.; data curation, S.A.D.; writing—original draft preparation, S.A.D., K.B.G.; writing—review and editing, S.A.D., P.K. and T.A. All authors have read and agreed to the published version of the manuscript.

Funding: This research received no external funding.

Conflicts of Interest: The authors declare no conflict of interest.

References

1. Quan, Z.; Cui, S.; Sayed, A.H. Optimal linear cooperation for spectrum sensing in cognitive radio networks. *IEEE J. Sel. Signal Process.* **2008**, *2*, 28–40. [CrossRef]
2. Niyato, D.; Hossain, E. Spectrum trading in cognitive radio networks: A market equilibrium based approach. *IEEE Wirel. Commun.* **2008**, *15*, 71–80. [CrossRef]
3. Nasif, A.O.; Mark, B.L. Opportunistic spectrum sharing with multiple cochannel primary transmitters. *IEEE Transact. Wirel. Commun.* **2009**, *8*, 11. [CrossRef]
4. Yue, G.; Wang, X. Anti-jamming coding techniques with application to cognitive radio. *IEEE Transact. Wirel. Commun.* **2009**, *8*, 12. [CrossRef]
5. Devaraj, S.A.; Aruna, T.; Muthukumaran, N.; Roobert, A.A. Adaptive cluster-based heuristic approach in cognitive radio networks for 5G applications. *Trans. Emerg. Tel. Tech.* **2022**, *33*, e4383. [CrossRef]
6. Rashid, M.M.; Hossain, J.; Hossain, E.; Bhargava, V.K. Opportunistic spectrum scheduling for multiuser cognitive radio: A queueing analysis. *IEEE Transact. Wirel. Commun.* **2009**, *8*, 5259–5269. [CrossRef]
7. Song, C.; Zhang, Q. Cooperative spectrum sensing with multi-channel coordination in cognitive radio networks. In Proceedings of the IEEE International Conference on Communications, Cape Town, South Africa, 23–27 May 2010; p. 1.
8. Wang, B.; Zhao, D.; Cai, J. Joint connection admission control and packet scheduling in a cognitive radio network with spectrum underlay. *IEEE Transact. Wirel. Commun.* **2011**, *10*, 3852–3863. [CrossRef]
9. Nguyen, M.V.; Lee, H.S. Effective scheduling in infrastructure-based cognitive radio networks. *IEEE Transact. Mobile Comput.* **2011**, *10*, 853–867. [CrossRef]
10. Azarfar, A.; Frigon, J.-F.; Sans'o, B. Analysis of cognitive radio networks based on a queueing model with server interruptions. In Proceedings of the IEEE International Conference on Communications, ICC, Ottawa, ON, Canada, 10–5 June 2012; pp. 1703–1708.
11. Li, X.; Li, W.; Hei, Y. Joint spectrum sensing and user selection strategy for cognitive radio networks. In Proceedings of the International Conference on Wireless Communications & Signal Processing (WCSP), Huangshan, China, 25–27 October 2012; pp. 1–6.
12. Ruby, D.; Vijayalakshmi, M. Scheduling scheme with dynamic packet shifting in multichannel cognitive radio adhoc network. In *Emerging Research in Computing, Information, Communication and Application*; Elsevier Publications: Amsterdam, The Netherlands, 2013.
13. Xue, D.; Ekici, E. Efficient distributed scheduling in cognitive radio networks in the many-channel regime. In Proceedings of the International Symposium and Workshops on Modeling and Optimization in Mobile, Ad Hoc and Wireless Networks, Tsukuba, Japan, 13–17 May 2013.
14. Ji, C.; Nie, X.; Yuan, Y. A cross-layer scheduling algorithm based on cognitive radio network. *Appl. Math. Inf. Sci. Int. J.* **2013**, *7*, 611–617. [CrossRef]
15. Yu, L.; Liu, C.; Zhu, W.; Hua, S.; Wang, W. Bandwidth efficient and rate-adaptive video delivery in TV white space. *IEEE Transact. Circuits Syst. Video Technol.* **2014**, *24*, 1605–1619. [CrossRef]
16. Parzy, M.; Bogucka, H. Coopetition methodology for resource sharing in distributed OFDM-based cognitive radio networks. *IEEE Transact. Commun.* **2014**, *62*, 1518–1529. [CrossRef]
17. Abbas, N.; Nasser, Y.; Ahmad, K.E. Recent advances on artificial intelligence and learning techniques in cognitive radio networks. *EURASIP J. Wirel. Commun. Netw.* **2015**, *2015*, 174. [CrossRef]
18. Tsakmalis, A.; Chatzinotas, S.; Ottersten, B. Centralized power control in cognitive radio networks using modulation and coding classification feedback. *arXiv* **2015**, preprint. arXiv:1510.06634. [CrossRef]
19. Zhu, X.; Yang, B.; Chen, C.; Xue, L.; Guan, X.; Wu, F. Crosslayer scheduling for OFDMA-based cognitive radio systems with delay and security constraints. *IEEE Transact. Veh. Technol.* **2016**, *64*, 5919–5934. [CrossRef]
20. Orumwense, E.F.; Afullo, J.T.; Srivastava, V.M. Achieving a better energy-efficient cognitive radio network. *Int J. Comput. Inf. Syst. Ind. Manag. Appl.* **2016**, *8*, 205–213.

21. Alberti, A.M.; Mazzer, D.; Bontempo, M.M.; de Oliveira, L.H.; da Rosa Righi, R.; Sodré, A.C. Cognitive radio in the context of internet of things using a novel future internet architecture called Nova Genesis. *Comput. Electr. Eng.* **2018**, *57*, 147–161. [CrossRef]
22. Heejung, Y.; Yousaf, B. Cognitive radio networks for internet of things and wireless sensor networks. *Sensors* **2020**, *20*, 5288. [CrossRef] [PubMed]
23. Li, X.; Zhou, R.; Zhou, T.; Liu, L.; Yu, K. Connectivity probability analysis for green cooperative cognitive vehicular networks. *IEEE Transact. Green Commun. Netw.* **2022**, *6*, 1553–1563. [CrossRef]
24. Sun, Y.; Yu, K.; Bashir Liao, X. Bl-IEA: A bit-level image encryption algorithm for cognitive services in intelligent transportation systems. *IEEE Transact. Intell. Transp. Syst.* **2021**, 1–13. [CrossRef]
25. Kottursamy, K.; Rehman Khan, A.; Sadayappillai, B.; Raja, G. Optimized D-RAN aware data retrieval for 5G information centric networks. *Wirel. Pers. Commun.* **2022**, *124*, 1011–1032. [CrossRef]

Article

A 0.6 V_{IN} 100 mV Dropout Capacitor-Less LDO with 220 nA I_Q for Energy Harvesting System

Yuting Zhang [1,2], Qianhui Ge [1,2] and Yanhan Zeng [1,2,*]

[1] School of Electronics and Communication Engineering, Guangzhou University, Guangzhou 510000, China; iyuting_2002@163.com (Y.Z.); 2112130007@e.gzhu.edu.cn (Q.G.)

[2] Key Lab of Si-Based Information Materials & Devices and Integrated Circuits Design, Guangzhou University, Guangzhou 510000, China

* Correspondence: yanhanzeng@gzhu.edu.cn

Abstract: A fully integrated and high-efficiency low-dropout regulator (LDO) with 100 mV dropout voltage and nA-level quiescent current for energy harvesting has been proposed and simulated in the 180 nm CMOS process in this paper. A bulk modulation without an extra amplifier is proposed, which decreases the threshold voltage, lowering the dropout voltage and supply voltage to 100 mV and 0.6 V, respectively. To ensure stability and realize low current consumption, adaptive power transistors are proposed to enable system tropology to alter between 2-stage and 3-stage. In addition, an adaptive bias with bounds is utilized in an attempt to improve the transient response. Simulation results demonstrate that the quiescent current is as low as 220 nA and the current efficiency reaches 99.958% in the full load condition, load regulation is 0.0059 mV/mA, line regulation is 0.4879 mV/V, and the optimal PSR is −51 dB.

Keywords: low-dropout regulator; ultra-low-input voltage; high power efficiency; bulk modulation; energy harvesting

Citation: Zhang, Y.; Ge, Q.; Zeng, Y. A 0.6 V_{IN} 100 mV Dropout Capacitor-Less LDO with 220 nA I_Q for Energy Harvesting System. *Micromachines* **2023**, *14*, 998. https://doi.org/10.3390/mi14050998

Academic Editors: Qiongfeng Shi and Huicong Liu

Received: 9 April 2023
Revised: 28 April 2023
Accepted: 2 May 2023
Published: 3 May 2023

1. Introduction

Energy harvesting (EH) technology converts weak energy harvested from the environment into electricity [1–3]. EH has been used in many fields, such as the Internet of Things (IoT), wearable devices, and wireless sensors [4–6]. The output voltage collected from energy harvesting is unstable. Therefore, the low-dropout regulators (LDOs) are used to provide a stable power supply for the loads, as they have the advantages of uncomplexity, a low cost, and being noise-immune.

A typical EH system is depicted in Figure 1. Considering the weakness of the energy source and low collection efficiency, energy collection suffers from a low supply voltage and high power consumption [7–9]. Consequently, new challenges and requirements are set forth for the supply voltage and power conversion efficiency of LDO.

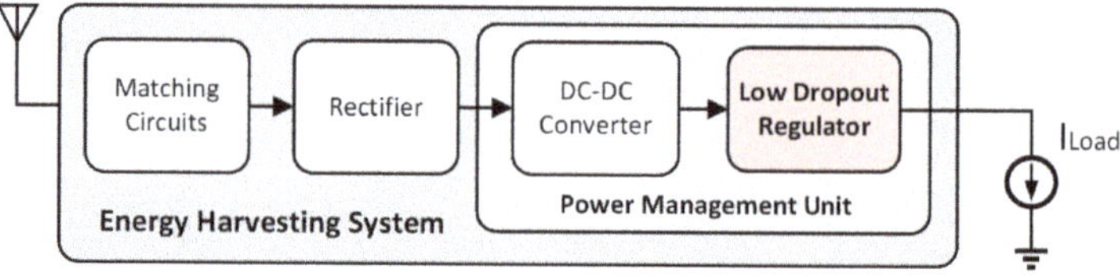

Figure 1. A typical EH system.

For the purpose of improving the integration and reducing the cost, [10–13] proposed the output-capacitor-less (OCL) LDOs consuming μA-level static currents. However, low-power equipment should consume as little electricity as possible during standby and operation. To further reduce the current to nA levels, [14] adopts an adaptive bias solution based on the super-source follower (SSF) structure. Nevertheless, the minimum supply voltage is as high as 1.8 V, and struggles to meet the EH requirement. In addition, the large dropout voltage leads to large power loss.

Another approach to improving efficiency is to reduce the dropout voltage [15]. A dropout voltage can be defined as the product of the valid channel resistance of the component and the load current in non-regulatory conditions [16]. In [17], a large and wide ratio power transistor is proposed to reduce the equivalent resistance, but this requires a large area. One method to reduce the threshold voltage is the floating meter [18,19], and the other method is body bias [20–22]. Recent research has shown that the threshold voltage can be reduced using current-driven bulks [23]. Since V_B is less than V_S in the PMOS power transistor, the source-end bipolar crystal pipe can be turned on to reduce the threshold voltage. This extra adaptively biased current-driven loop generates a lot of static current, which cannot meet the low I_Q requirements.

This paper proposes a 100 mV-dropout and high-efficiency OCL-LDO with low supply voltage, which uses bulk modulation and adaptive bias technology with adaptive power transistors (APT). The bulk modulation without using the auxiliary amplifier has significantly reduced the dropout voltage to 100 mV. The adaptive bias can quickly detect the transient jumps to shorten the recovery times and reduce the quiescent current. Through the use of APT, the system is able to switch between 2-stage and 3-stage amplifiers under different load conditions, ensuring its stability in the full load range. Further, APT reduces power consumption by reducing unnecessary current consumption. On the one hand, ultra-low V_{IN} and V_{OUT} are suitable for weak EH systems, since they require a low input voltage and output voltage. On the other hand, by reducing the dropout voltage, high efficiency can be achieved with low power consumption.

In the rest of this paper, the operating principle of the proposed LDO is discussed in Section 2. Sections 3 and 4 describe the circuit implementation and stability analysis, respectively. The simulation results, along with comparisons with the state-of-the-art designs, are reported in Section 5. Finally, conclusions are drawn in Section 6.

2. Operation Principle

2.1. Bulk Modulation without Amplifier

The amount of energy collected by the EH system is extremely small. DC-DC Boost converter only produces a few hundred mV of output voltage, requiring a low V_{IN} in the LDO design. Generally, the V_{IN} can be expressed as:

$$V_{IN} = V_G + |V_{GSP}| \geq V_G + |V_{THP}|, \tag{1}$$

where V_G, V_{GSP}, and V_{THP} are the gate voltage, gate-source voltage, and threshold voltage of the power transistor, respectively. When there is a voltage difference between the bulk and source, V_{TH} decreases with V_{BS} due to the body effect and can be written as:

$$V_{TH} = V_{TH0} + \gamma(\sqrt{|2\Phi_F| - V_{BS}} - \sqrt{|2\Phi_F|}), \tag{2}$$

where V_{TH0} is the threshold voltage when V_{BS} is zero, γ denotes the bulk-effect coefficient, and Φ_F is the Fermi potential. Generally, γ ranges from 0.3 $V^{1/2}$ to 0.4 $V^{1/2}$. This analysis was simplified by replacing source-bulk voltage V_{SB} with $-V_{BS}$. In the case of $a^2 >> b^2$, taking into account a mathematical formula [24]:

$$\sqrt{a + b} \approx \sqrt{a} + \frac{b}{2\sqrt{a}}. \tag{3}$$

In Equation (2), there is $(2\Phi_F)^2 >> V_{SB}^2$, and substitute them into a and b, and considering the relationship of [25,26]:

$$1 + \frac{\gamma}{2\sqrt{2\Phi_F}} = \eta,\tag{4}$$

where η ranges from 1.2 to 1.3. Part of Equation (2) can be approximately expressed as follows:

$$\gamma\left(\sqrt{2\Phi_F - V_{BS}} - \sqrt{2\Phi_F}\right) \approx \frac{\gamma V_{SB}}{2\sqrt{2\Phi_F}} = (\eta - 1)V_{SB}.\tag{5}$$

Therefore, we can conclude that V_{TH} is linearly related to V_{SB}. Via the body effect approximation, Equation (2) can be simplified as [27]:

$$V_{TH} \approx V_{TH0} + (\eta - 1)V_B.\tag{6}$$

Previously, the bulk of the power transistor was connected to the source to prevent the conduction of the PN junction between the source and the bulk, as shown in Figure 2a. Thus $V_{TH} = V_{TH0}$. The minimum V_{IN} of this kind of LDO is larger than 1 V, which does not meet the criteria for low V_{IN}.

The bulk modulation is proposed to reduce V_{TH} and then V_{IN}, as illustrated in Figure 2b. This system imposes an additional amplifier to generate the bulk voltage V_{BODY}, which increases power consumption.

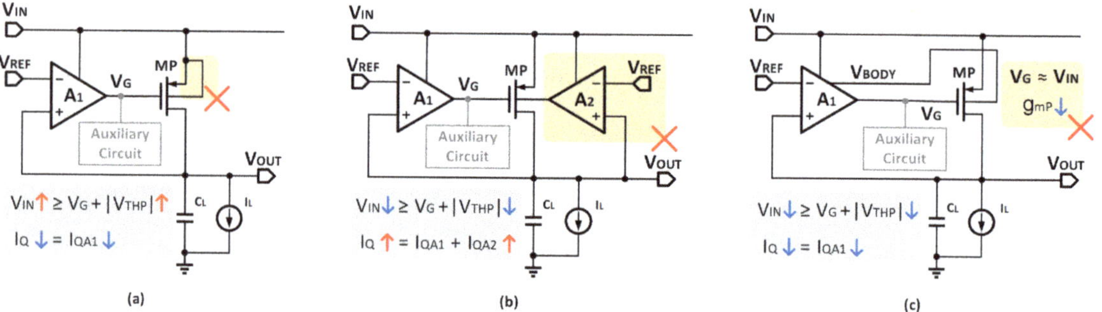

Figure 2. Structures of (**a**) conventional analog LDO, (**b**) bulk modulation with an extra amplifier, (**c**) bulk modulation without an extra amplifier.

Recently, bulk modulation without an extra amplifier was proposed in [28], which can better realize the balance between power consumption and low V_{IN}. Figure 2c shows its structural block diagram. However, the modulated V_G is close to V_{IN}, resulting in low gain and poor performance.

To solve the problems mentioned above, this paper proposes a voltage reuse scheme to generate V_{BODY}. In a typical three-stage structure of OCL-LDO, the buffer is used to increase the drive ability and enhance the load capacity of the power transistor due to the small output resistance. The buffer output is usually only connected to the gate of the power transistor V_G. In this paper, the bulk modulation is simplicity obtained by connecting the buffer output to not only the gate but also the bulk of the power transistor, as shown in Figure 3. V_{IN} in this paper can be expressed as:

$$V_{IN} = V_G + |V_{TH}| = V_{TH0} + V_G + (\eta - 1)V_G = V_{TH0} + \eta V_G.\tag{7}$$

From Equation (7) V_{IN} is determined only by V_{TH0} and V_G. The large value of V_{BODY} is replaced by V_G, resulting in a reduction in V_{IN} to 600 mV. In addition, a gain path is established from the bulk to V_{OUT}, and the body transconductance g_{mb} can enhance the regulation gain. Consequently, V_{IN} is more effectively reduced without introducing additional power consumption, ensuring that the transistor performs normally in low V_{IN}.

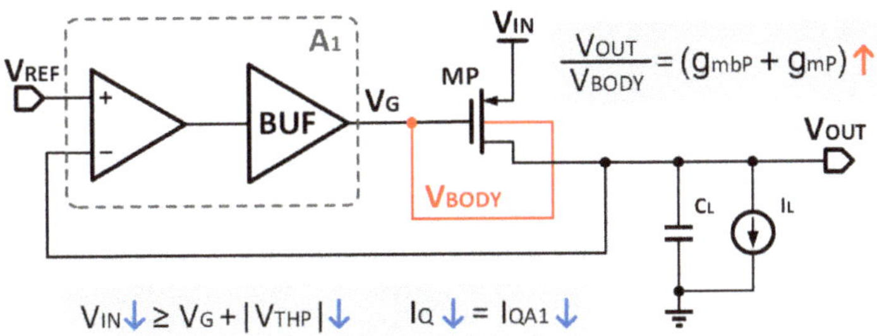

Figure 3. The structural diagram of the V_{BODY} reuse.

It is important to note that a forward-biased PN junction between bulk and source experiences very low voltage. An exceedingly high V_{SB} may lead to the forward breakdown of the MOS transistor, resulting in device damage. Lowering the bulk voltage appropriately, while ensuring that the PN junction does not undergo breakdown, can reduce V_{TH}. To validate the feasibility of the proposed approach, simulations of the circuit were conducted. Figure 4a shows that the value of V_{TH} as V_{IN} varies under different V_{BODY} when adopting the V_{BODY} reuse method. The simulation results show that V_{TH} increases as V_{IN} decreases, while V_{BODY} remains constant. However, V_{TH} can be effectively reduced by decreasing V_{BODY}. In the proposed design, the MOS transistor is protected from forward breakdown since the maximum value of V_{BODY} is 280 mV, which is less than the forward breakdown voltage. Additionally, Figure 4b shows the drain current with and without V_{BODY} reusing. There is evidence that, within a specific range, after adopting the reuse of V_{BODY}, I_Q can be effectively reduced. Therefore, the proposed bulk modulation without an amplifier reduces both V_{IN} and I_Q.

Despite this, bulk modulation also has some disadvantages, such as circuit instability due to the excessive modulation or reverse breakdown of PN junctions caused by a V_{BODY} that is too low. To solve the above problems, APT technology is proposed in the next section.

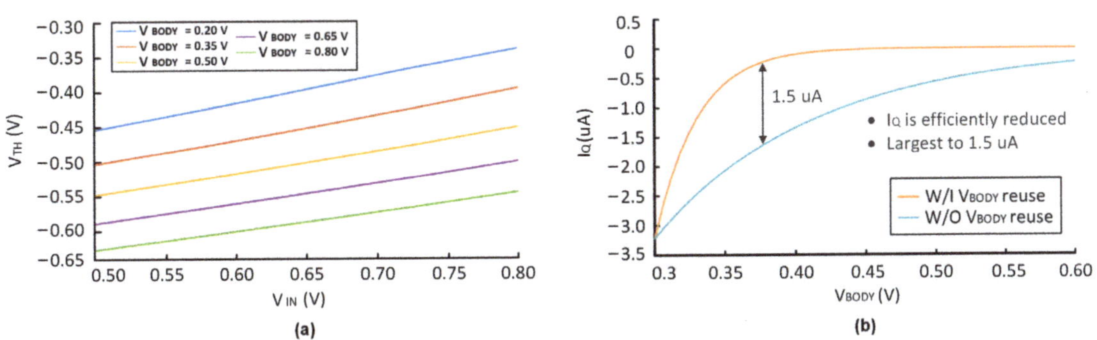

Figure 4. Simulation results of (**a**) the change of V_{TH} with V_{IN}, in the case of different V_{BODY}, (**b**) the I_Q with V_{BODY} reuse and without V_{BODY} reuse.

2.2. Adaptive Power Transistors

The drain current I_D of the power transistor working in the saturation region can be given by:

$$I_D = \frac{1}{2}\mu_P C_{OX}\frac{W}{L}(|V_{GS}| - |V_{TH}|)^2, \tag{8}$$

where μ_P is the mobility of the PMOS channel, C_{OX} is the capacitance of the gate oxide layer, and W/L is the transistor ratio.

As the load current I_{LOAD} gradually increases and V_{GS} increases, so V_G decreases accordingly. When employing V_{BODY} reuse, there is $V_G = V_{BODY}$. Reducing V_{BODY} too much will result in increased V_{BS}, which may lead to the forward breakdown of the PN junction.

As a preventative measure, this paper proposes the APT technology. This is based on the principle that, when the load changes, a feedback signal is used to adjust the use of the auxiliary and main power transistors (MP_1 and MP_2). Meanwhile, bulk modulation performs on MP_1 and MP_2. Specifically, the modulation effect will be enhanced as I_{LOAD} increases. The V_{BODY} reuse scheme being implemented for MP_2 will turn off under light loads, making V_{BODY} close to V_{IN}. The schematic diagram and working process of APT are shown in Figure 5.

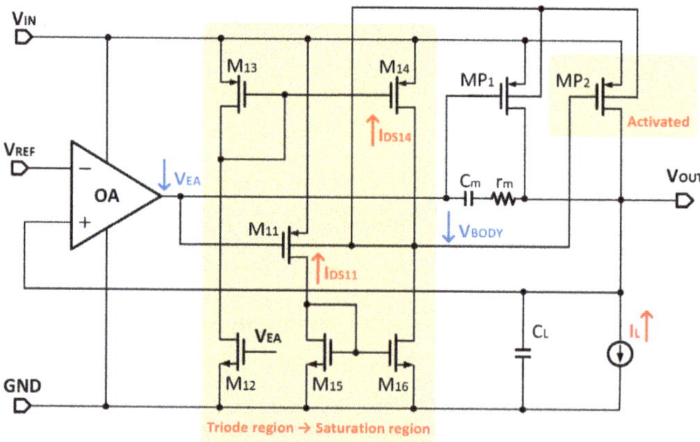

Figure 5. The schematic diagram and working process of APT.

The system works in a twp-stage amplifier topology when the load is light. Due to the second gain stage performing in the triode region, M_{14} pulls the output voltage closer to the supply. Therefore, the V_{GS} of MP_2 is minimal, and MP_2 is turned off. This was obtained by designing the aspect ratio of M_{14} and M_{16}. M_{14} is set to be small enough and M_{16} is set to be large enough; thus, M_{16} works in the saturation region, whereas M_{14} is forced into the deep linear region to sustain equilibrium in the light load. The gate of MP_1 is connected to the output of the first-stage amplifier, V_{EA}. MP_2 is forcibly disabled and all the load current flows through MP_1.

When the load current increases, the I_{DS} of MP_1 increases, as well as its V_{GS}. Therefore, V_{EA} decreases, as well as the V_G of M_{11}, which increases the current of M_{11} and provides a large current to M_{15}, as well as M_{16}. Thus, the drop-down ability of M_{16} becomes strong enough to pull M_{14} from the linear region into the saturation region. Hence, MP_2 is activated. The system works in a 3-stage amplifier topology.

To better distinguish the two operating states, we define a threshold current I_{ON} of approximately 350 μA. When the load current is less than I_{ON}, the second-stage amplifier is working in the triode region and turns off MP_2. When the load current increases to I_{ON}, the second-stage amplifier works in the saturation region and turns on MP_2.

It is worth mentioning that the proposed LDO can adaptively switch the power transistors according to the load current, allowing the system to transform itself between 2-stage and 3-stage topologies. In addition, simulation results have shown that V_{BODY} ranges between 350 mV and 550 mV with APT, and V_{BS} is less than the forward breakdown voltage, which ensures the safety of the device.

2.3. Structure of Proposed OCL-LDO

Figure 6 shows the LDO structure with the proposed bulk modulation and APT. The gain boost stage with an adaptive bias module (ABM) is used, which obtains a large bandwidth gain depending on the load current. The unity gain buffer (BUF) produces V_{BODY} to modulate the bulk voltages of MP_1 and MP_2, lowering the total current consumption via bulk modulation. The APT not only achieves satisfactory linearity but also improves the transient response. In an attempt to realize enough stability, the compensation module (CM) with an internal RC network is used.

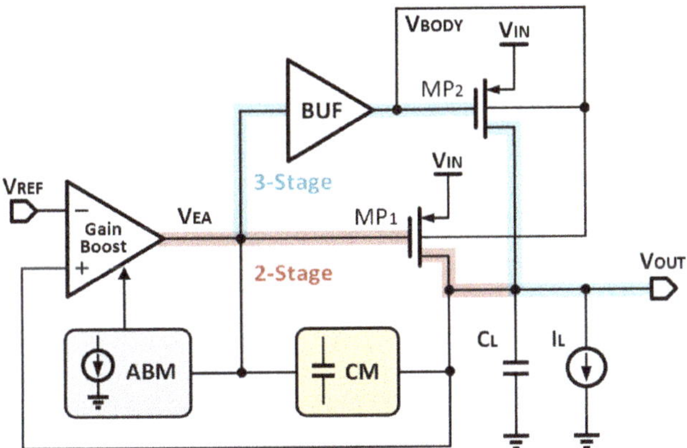

Figure 6. The black box schematic of the proposed LDO.

3. Circuit Implementation

This section will provide an overview of the OCL-LDO, including its structural realization and design consideration. The proposed OCL-LDO works under the condition of $V_{IN} = 0.6$ V, $V_{REF} = 0.5$ V, with $V_{Drop} = 100$ mV. The LDO's V_{REF} is ultra-low, which can be generated with a BJT-based Kuijk bandgap reference [29]. Figure 7 depicts the complete circuit diagram. The gain boost stage is a transconductance-boosted amplifier (M_1–M_{10}) and ABM provides adaptive tail current. The BUF consists of M_{11}–M_{16} in the form of a current mirror, which is capable of increasing the slew rate and improving the transient response. The CM is composed of a resistor r_m and a capacitor C_m.

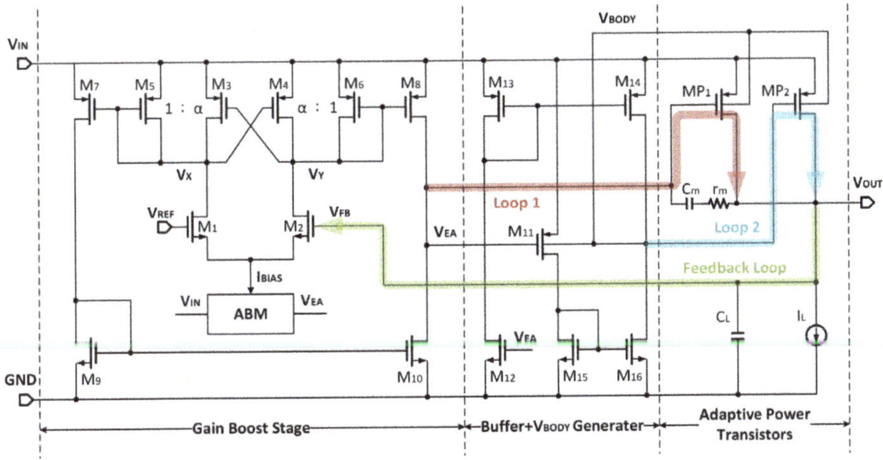

Figure 7. The schematic of the proposed LDO.

3.1. Maximize the Efficiency under Low V_{IN}

To maximize efficiency, it is necessary to reduce the dropout voltage. However, lowering the dropout voltage degenerates the DC gain of the power stage, which can be expressed as:

$$
\begin{aligned}
A_P &= g_{m,linear} R_{OUT} \\
&\approx \frac{\mu_n C_{OX} \frac{W}{L} V_{DS}}{\mu_n C_{OX} \frac{W}{L} (V_{GS} - V_{TH})} \\
&= \frac{V_{DS}}{V_{GS} - V_{TH}},
\end{aligned}
\tag{9}
$$

where $g_{m,linear}$ is the transconductance in the linear region. As V_{DS} in the design is merely 100 mV; it is highly possible for it to be smaller than $(V_{GS} - V_{TH})$, implying that the A_P is less than one. It is worth noting that this issue is not present for regular 200 mV dropout regulators, as their power stage can supply roughly 20 dB gain for the control loop. To address this problem, the proposed LDO introduces a gain boost stage with ABM.

From Figure 7, the gain boost stage employs a single-stage differential amplifier along with a cross-coupled PMOS pair to increase the gain. The cross-coupled pair generates negative resistance that can counteract the equivalent resistance of diode-connected transistors M_5 and M_6, which effectively increases the resistance of internal nodes V_X and V_Y. The amplifier gain is obtained as:

$$
A_{EA} = \frac{(1+\alpha)g_m + g_{ds}}{(1-\alpha)g_m + g_{ds}} g_{m1}(r_{o8} \| r_{o10}),
\tag{10}
$$

where g_m represents the transconductance of M_3, M_4, and g_{ds} denotes the total output conductance of node V_X or V_Y. The gain is amplified by a factor of $((1+\alpha)g_m + g_{ds})/((1-\alpha)g_m + g_{ds})$. To ensure that it works as an amplifier rather than a hysteresis comparator, the pairs M_3-M_4 and M_6-M_5 are set at a ratio of α:1, where α should be kept smaller than 1 to avoid the over-compensation of negative resistance, taking into account the random mismatch under the fabrication.

For the push-pull stage structure, the DC gain can be approximated at 1. As a result, the overall circuit gain can be expressed as follows:

$$
A_{DC} = A_P \cdot A_{EA} = \frac{V_{DS}}{V_{GS} - V_{TH}} \cdot \frac{(1+\alpha)g_m + g_{ds}}{(1-\alpha)g_m + g_{ds}} g_{m1}(r_{o8} \| r_{o10}).
\tag{11}
$$

Thus, the DC performance of this LDO can be significantly improved by increasing A_{EA} when A_P is small.

3.2. Achieve Low I_Q and Transient Response

Realizing nA-level quiescent current is necessary to ensure ultra-high-power efficiency. The proposed LDO uses adaptive bias to ensure a low quiescent current while reducing transient response time and improving transient response.

Unlike traditional adaptive bias circuits, which mirror the current at the gate of the power transistor to bias the amplifier or buffer, the proposed ABM is driven by the first-stage amplifier itself and provides a bias current proportional to the load current, which can maintain a large enough gain of the first-stage amplifier over the entire load range.

Figure 8 illustrates the specific design of ABM. The ABM's input current is set by M_{23}, which is biased by V_{EA} and is proportional to the load current because V_{EA} is closely related to the load current. The output current of ABM is I_{BIAS}, which is used to bias the amplifier and equal to the sum current of M_1 and M_2. I_{MIN} and I_2 are used to define the bias upper and lower limit. The workflow of ABM can be summarized into the following three phases depending on the load current.

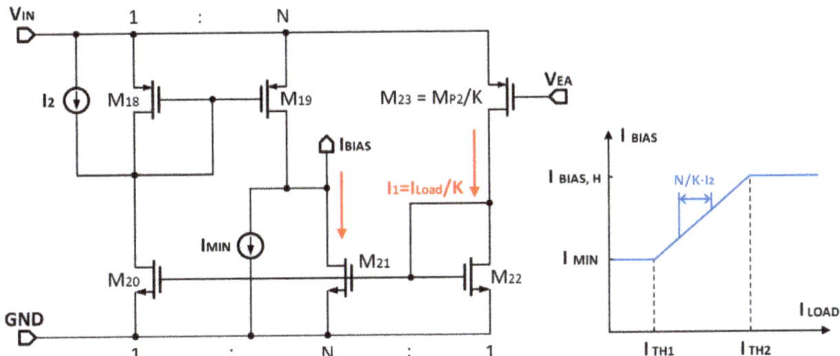

Figure 8. The schematic of ABM.

- Light load

During periods of low load current, the current in M_{23} is ultra-low and not sufficient to turn on the transistor. In this case, all devices are turned off, and I_{BIAS} is equal to I_{MIN}.

$$I_{BIAS,L} = I_{MIN}. \tag{12}$$

- Moderate load

As the load current increases, the transistors M_{20}, M_{21} and M_{22} in the ABM are turned on, and there are

$$I_{DM20} = \frac{1}{K} \cdot I_{LOAD},$$
$$I_{DM21} = \frac{N}{K} \cdot I_{LOAD}. \tag{13}$$

Considering that the current of M_{20} is less than I_2, I_{BIAS} is equal to the current sum of M_{21} and I_{MIN}, and can be expressed as:

$$I_{BIAS,M} = I_{MIN} + I_{DM21} = I_{MIN} + \frac{N}{K} \cdot I_{LOAD}, \tag{14}$$

where K and N are the current mirror ratios, as shown in Figure 8.

- Heavy load

Once the current of M_{20} increases to larger than I_2, the current mirror, made up of M_{18} and M_{19}, is activated. This threshold current I_{TH2} can be expressed in the following manner:

$$I_{LOAD} = I_{TH2} = K \cdot I_2. \tag{15}$$

I_{BIAS} stops increasing and reaches the upper limit, which can be expressed as:

$$I_{BIAS,H} = I_{MIN} + \frac{N}{K} \cdot I_{LOAD} = I_{MIN} + N \cdot I_2, \tag{16}$$

The bounds of I_{BIAS} ensures stable operation. In the actual design, I_{MIN} and I_2 are set at 7 nA, 18 nA, respectively. N and K are set at 6 and 5150, respectively. It is worth mentioning that ABM also contributes to improvements in gain. The equivalent transconductance g_{mEA} of the gain boost stage can be determined as follows:

$$g_{mEA} = \frac{g_{m1}}{\frac{g_{m3}}{g_{m5}} - 1}, \tag{17}$$

As I_{LOAD} increases, g_{m1} increases significantly; thus, g_{mEA} enlarges accordingly. Therefore, the bias current for the amplifier is proportional to the load current.

4. Stability Analysis

To analyze the stability of the proposed LDO, a small-signal transfer function needs to be derived. It can be seen from Figure 9 that the small-signal model switches between the two-stage amplifier and the three-stage amplifier. Due to the existence of the self-adaptive power transistor, MP_1, and MP_2 turn on and work sequentially under the condition of different load currents. Thus, the stability analysis of the circuit should be discussed according to the situation: (i) the second-stage amplifier and (ii) the third-stage amplifier.

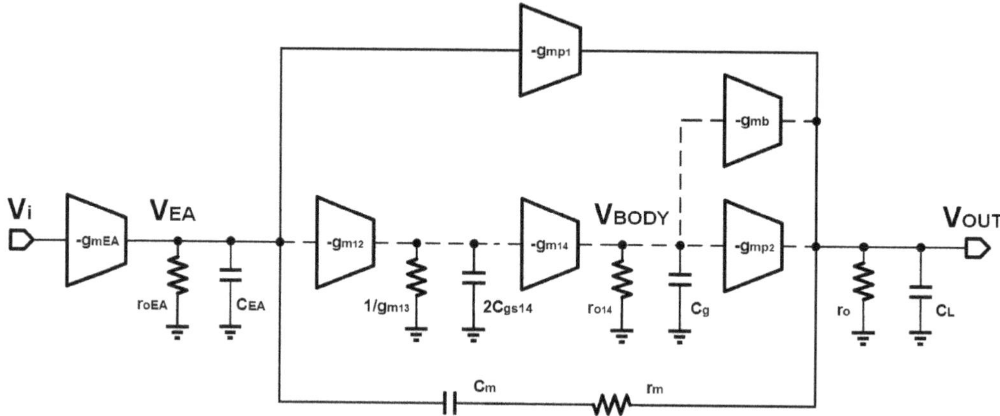

Figure 9. The small-signal model of the proposed LDO.

Due to the topology complexity, it is difficult to obtain the transmission function and analyze the stability. To simplify the derivation of the loop without reducing accuracy, the following assumptions are established.

The gains of the first-stage amplifier, the main power transistor, and the side power transistor are far larger than 1, that is, $g_{m<i>}r_{o<i>} \gg 1$. In the case of the third-stage amplifier, $g_{mp2} \gg g_{mp1}$. Since the g_{mb} of MP_1 is small, it was ignored in the analysis. The output capacitor C_L is larger than compensating capacitors C_M, C_{EA}.

4.1. Two-Stage Structure

In the case of a small load current, the LDO is equivalent to a secondary structure. There is a pole in V_{OUT} due to the output resistance and load capacitance. Since output resistance is closely related to the load size, a light load will have a relatively large output resistance, which results in the pole associated with the output terminal moving to the low frequency and becoming the dominant pole, P_1. The secondary pole P_2 is located at the output of the first stage because of the output resistance of the amplifier and the parasitic capacitance of the MP_1 gate. The poles' frequencies can be obtained by calculation as follows:

$$P_1 = -\frac{1}{r_{oEA}C_m g_{mp1} r_o},\tag{18}$$

$$P_2 = -\frac{g_{mp1}}{C_{EA}}.\tag{19}$$

Maintaining stability by only allowing one pole to exist within the bandwidth is necessary. When the OCL-LDO is operated under the minimum load current condition, the pole frequency located in the output is close to zero. Therefore, LDO has a very poor phase margin (PM) and must adopt a compensation structure to produce a zero point. The compensation is performed by connecting the compensation Miller capacitor C_M between the output of the first-stage amplifier and V_{OUT}. However, the traditional Miller compensation produces a zero point in the right half plane (RHP), which is converted to the left half plane (LHP) by adding the zeroing resistor r_m. Therefore, the zero point Z_1 can be expressed as:

$$Z_1 = -\frac{g_{mp1}}{C_m\left(g_{mp1}r_m - 1\right)}. \tag{20}$$

Therefore, the transfer function of the two-stage structure can be expressed as:

$$LG(s) = A_0\frac{\left(1 + \frac{s}{z_1}\right)}{\left(1 + \frac{s}{p_1}\right)\left(1 + \frac{s}{p_2}\right)}, \tag{21}$$

where A_0 is the DC gain under the condition of light load and can be calculated as:

$$A_0 = g_{mEA}r_{oEA}g_{mp1}r_o. \tag{22}$$

As shown in Figure 10, the LDO's open-loop frequency response and pole-zero distribution are plotted under the conditions of V_{IN} = 0.6 V with I_{LOAD} = 10 μA. Based on the simulation results, the PM is better than 45° for the worst-case scenario with the smallest load current, and the overall circuit is stable.

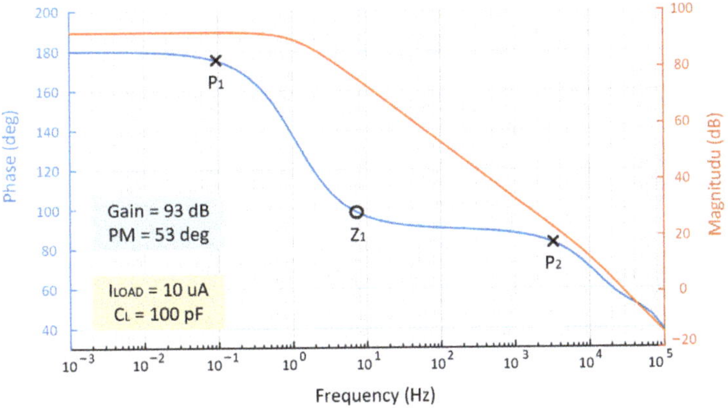

Figure 10. Open-loop gain simulation of the proposed LDO when I_{LOAD} = 10 μA.

4.2. Three-Stage Structure

During heavy load conditions, the output resistance of the circuit decreases, and the pole at the output becomes the secondary pole P_2 instead of the dominant pole P_1. Since the MP_2 main power transistor has a large size, its parasitic capacitance is relatively large. Therefore, the pole existing on the gate of the main power transistor becomes the main pole P_1. The poles under heavy load can be approximated as follows:

$$P_1 = -\frac{g_{m13}}{C_m g_{m12}g_{m14}r_{oEA}r_{o14}}, \tag{23}$$

$$P_2 = -\frac{1}{C_L\left(r_{mp2}\|r_o\right)}, \tag{24}$$

$$P_3 = -\frac{g_{m13}}{2C_{gs14}}, \tag{25}$$

$$P_4 = -\frac{1}{C_g r_{o14}}. \tag{26}$$

Because the output capacitance and equivalent resistance of other nodes in the small-signal diagram are small, other poles are outside the bandwidth, which will be canceled out by zeros and thus have no effect on the stability of the overall circuit.

Due to the relationship between the equivalent capacitance of the miller compensation and the gain of the amplifier, the zero frequency changes in response and moves to a higher frequency. This zero is given by:

$$Z_1 = -\frac{g_{mp1}}{C_m}, \tag{27}$$

$$Z_2 = -\frac{g_{m12}g_{m14}\left(g_{mb} + g_{mp2}\right)}{C_m g_{mp1}}. \tag{28}$$

As a consequence, the three-stage structure has the following transfer function:

$$LG(s) = A_0 \frac{\left(1 + \frac{s}{z_1}\right)\left(1 + \frac{s}{z_2}\right)}{\left(1 + \frac{s}{p_1}\right)\left(1 + \frac{s}{p_2}\right)\left(1 + \frac{s}{p_3}\right)\left(1 + \frac{s}{p_4}\right)}, \tag{29}$$

where A_0 is the DC gain under the condition of heavy load, which can be calculated as:

$$A_0 = \frac{g_{mEA}g_{m12}g_{m14}\left(g_{mp2} + g_{mb}\right)r_{oEA}r_{o14}r_o}{g_{m13}}. \tag{30}$$

Figure 11 depicts the specific poles and zeros with a bandwidth under the heavy load. The figure shows that the PM is 59°, which is greater than 45°, indicating that the circuit is stable overall.

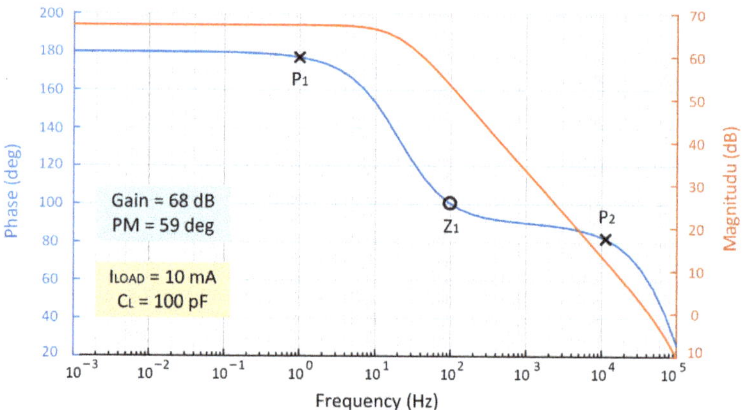

Figure 11. Open-loop gain simulation of the proposed LDO when I_{LOAD} = 10 mA.

5. Simulation Results and Discussion

The LDO proposed in this paper was implemented and verified in a 180 nm CMOS process, which supports 0.6–1.2 V supply voltage and 100 mV dropout voltage with a 10 mA maximum load current. The system can remain stable when C_L ranges between 0 and 1000 pF with a 10 pF on-chip compensation capacitor. When I_{LOAD} = 10 μA, V_{IN} varies from 0.6 V to 1.2 V, and I_Q changes from 226 nA to 219 nA. The devices' size and the most important parameter values for the proposed OCL-LDO are summarized in Table 1.

Table 1. Devices' size and parameters' value of proposed OCL-LDO.

Component	Size	Component	Size	Parameters	Value
M_1, M_2	5 μ/1 μ	M_{12}	500 n/5 μ	α	0.75
M_3, M_4	3 μ/1 μ	M_{13}	500 n/500 n	N	6
M_5, M_6, M_7	4 μ/1 μ	M_{14}	3.5 μ/500 n	K	5150
M_8, M_9	2 μ/1 μ	M_{16}	250 n/1 μ	I_{MIN}	7 nA
M_{10}, M_{15}	1 μ/1 μ	M_{P1}	70 μ/180 n	I_2	18 nA
M_{11}	1 μ/2 μ	M_{P2}	20 m/180 n		

Under the condition of V_{IN} = 600 mV, I_{LOAD} increases from 10 μA to 10 mA, I_Q changes from 220 nA to 1.2 μA. The changing trend of I_Q in the two cases is depicted in Figure 12.

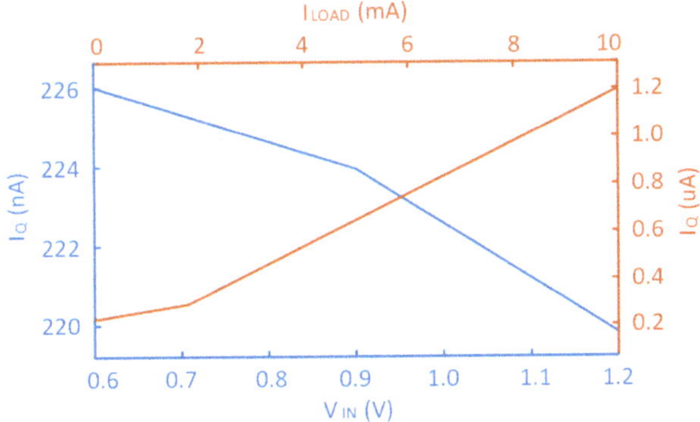

Figure 12. The I_Q of the proposed LDO when V_{IN} varies from 0.6 V to 1.2 V, I_{LOAD} changes from 10 μA to 10 mA.

Only 1.2 μA of quiescent current is consumed when the load current is 10 mA; thus, it achieves a maximum current efficiency of 99.96% when the load current is 10 mA. The current efficiency of the LDO under full load conditions is presented in Figure 13.

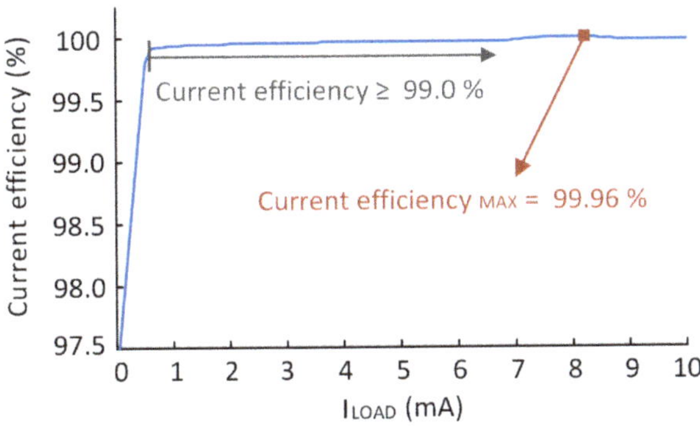

Figure 13. The current efficiency of the proposed LDO when I_{LOAD} changes from 10 μA to 10 mA.

The current of MP_1 and MP_2 are also shown in Figure 14. From the result, when the system works as a two-stage, system the current of MP_2 is almost equal to zero and MP_1 increases with I_{LOAD}. Once $I_{LOAD} > I_{ON}$, MP_2 is activated, its current increases with I_{LOAD} while the current of MP_1 remains unchanged.

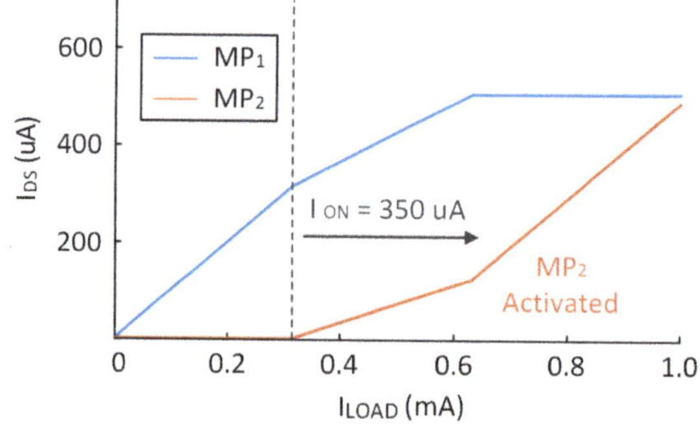

Figure 14. I_D of MP_1 and MP_2 when I_{LOAD} changes from 0 to 1 mA.

Additionally, we simulated the relationship between PM and load current under a variety of loop conditions. As shown in Table 2, PM decreases with an increase in I_{LOAD} when LDO works in a two-stage manner. However, PM in the three-stage is the opposite.

Table 2. PM with respect to I_{LOAD} in the two working conditions.

Loop	2-Stage			3-Stage		
I_{LOAD} (A)	100 µ	200 µ	300 µ	1 m	5 m	9 m
PM (deg)	49.1	47.8	44.5	51.1	55.4	58.2

The measured DC load regulation (LDR) is shown in Figure 15a; when the load current varies from 10 µA to 10 mA, the output voltage changes by 0.05857 mV. Therefore, the LDR of the LDO is 0.00585759 mV/mA. The simulation results depicted in Figure 15b indicate that the output voltage only changes to 0.292787 mV when the supply voltage changes linearly from 0.6 V to 1.2 V. According to the calculation formula, the DC line regulation (LNR) of LDO is 0.487978 mV/V.

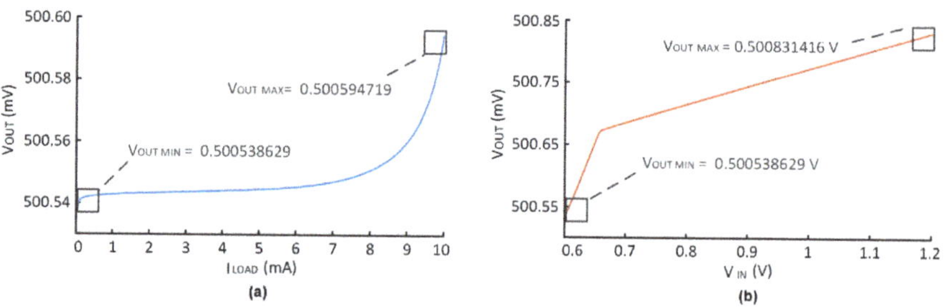

Figure 15. Simulation results of (**a**) DC load regulation, (**b**) DC line regulation.

Figure 16a,b show the measured transient response when the OCL-LDO supply voltage is set to 0.6 V and the load capacitor is set to 100 pF, respectively. For a 1 ns edge, the maximum transient response of $\triangle V_{OUT}$ is 230 mV, and the settling time is 5 µs. When the edge is 1 µs, the maximum $\triangle V_{OUT}$ is 140 mV, and the settling time is 3.5 µs, which represents a significant improvement. There is a minimum overshoot voltage of 95 mV and a undershoot voltage of 140 mV. However, the transient response is not entirely satisfactory. These results indicate that the circuit's performance is satisfactory and there is room for further improvement. Typically, an undershoot improvement circuit would be included to enhance the transient response. Nevertheless, due to the constraints of quiescent current, no improvement circuit has been incorporated.

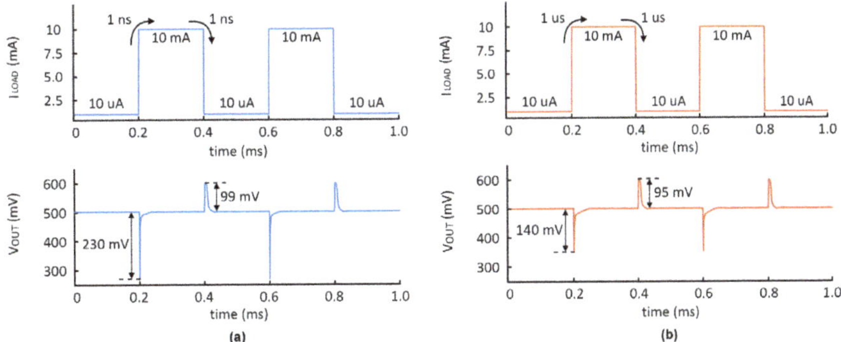

Figure 16. Measured dynamic response with a (**a**) 10 μA to 10 mA load step and 1 ns rising/falling edges, (**b**) 10 μA to 10 mA load step and 1 μs rising/falling edges.

The trend of the PSR with frequency can be plotted by analyzing the PSR at multiple frequencies over the course of one measurement. The simulated PSR performance at 10 μA load current, 0-pF C_L, and 100 mV dropout is shown in Figure 17. Hence, it can be seen that the minimum PSR is −51.40 dB, −49.17 dB at 1 Hz, and −43.97 dB at 1 k Hz.

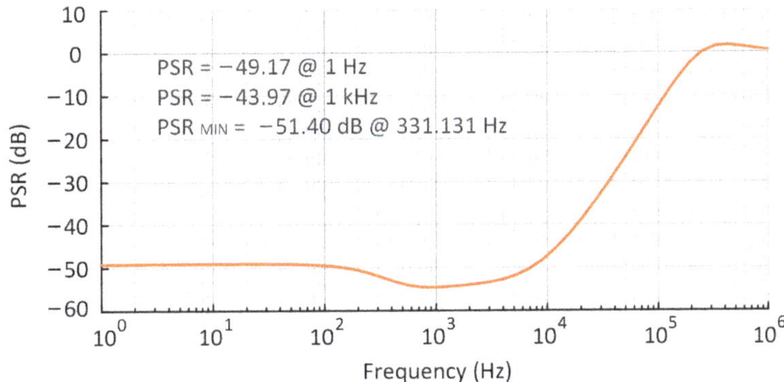

Figure 17. The simulated PSR performance of the proposed LDO at 10 μA load current, 0 pF C_L, and 100 mV dropout.

Simulations at different temperatures and process corners are conducted to demonstrate the robustness of the proposed design. The performance results, including minimum values of PM and GM, quiescent current, LDR, LNR, transient response, current efficiency, and PSR, are presented in Table 3. The minimum GM under different corners is greater than 50 dB, ensuring high regulation performance under different conditions. I_Q is generally within 500 nA, with the lowest value being 139 nA at the SS corner of −20 °C; the current efficiency is at least 99.95%, and the maximum is 99.99%. There is an excellent transient response at the FF corner of 27 °C, a 2.7 μs recovery time, and a maximum undershoot voltage of 320 mV. The worst PSR within 50 kHz is −51.4 dB at the TT corner of 27 °C. This shows that the proposed design can work steadily and perform as expected, even under extreme conditions. These results predict that the proposed LDO will work reliably after fabrication, even in the worst-case temperature and process corner.

Table 3. Performance summary under process and temperature corners.

Parameter	27 °C		−20 °C	80 °C	
Corner	TT	FF	SS	FF	SF
$Gain_{MIN}$ (dB)	68.4	65.1	78.51	71.1	53.4
PM_{MIN} (deg)	53.64	54.98	64.38	53.85	63.33
I_Q (nA)	220	370	139	870	424
LNR (mV/V)	0.48	0.78	0.60	0.83	0.73
LDR (μV/mA)	5.85	10.21	15.87	48.87	152.31
$\triangle V_{OUT}$ (mV)	230	320	346	465	428
Settling Time (μs)	3.8	2.7	7.1	5.3	13.1
Current Efficiency (%)	99.96	99.99	99.99	99.95	99.95
Min. PSR from DC to 50 kHz (dB)	−51.4	−63.4	−63.7	−55.9	−62.2

A merit map (*FoM*) is introduced to reflect the overall performance of the LDO, which has the following formula:

$$FoM = K\left(\frac{C_L I_Q \Delta V_{OUT}}{\Delta I_{O,MAX}^2}\right), \tag{31}$$

where I_Q is the minimum quiescent current, and ΔV_{OUT} is the maximum change in output during transient. Furthermore, K is the edge time ratio and defined by:

$$K = \frac{\Delta t \text{ used in the measurement}}{\text{the smallest } \Delta t \text{ among designs for comparison}}. \tag{32}$$

In this design, $K = 1$, bringing the above values into the calculation, the FoM value is 0.0506 ps. The comparison of the proposed LDO's performance with several state-of-the-art fully integrated LDOs is presented in Table 4 below. Based on the results of the analysis, the proposed LDO achieves the lowest dropout voltage, good load regulation, and lowest FoM.

Table 4. Performance comparison with other works.

Paper	TCAS-I [30]	TPE [31]	TPE [32]	AEU [33]	MDPI [11]	MDPI [10]	This Work
Year	2018	2018	2020	2021	2022	2022	2023
Technology (nm)	65	130	65	180	40	40	180
V_{IN} (V)	1	1	0.95	1.3	1.1	1.1	0.6
V_{OUT} (V)	0.8	0.8	0.8	1.1	0.9	0.9	0.5
Vdrop (mV)	200	200	150	200	200	200	100
Capacitor-less	Yes	Yes	Yes	Yes	Yes	Yes	Yes
$I_{LOAD(max)}$ (mA)	25	100	100	100	100	100	10
C_L (pF)	0–25 p	0–25 p	0–100 p	0–100 p	0–100 p	0–100 p	0–1000 p
I_Q (μA)	24.2	112	14	50.25	24.6-65	30	0.22
Line Reg. (mV/V)	0.7	2.25	12	0.75	N.A.	0.2	0.487
Load Reg. (mV/mA)	0.28	0.173	0.09	0.48	0.017	0.25	0.00585
Edge time (ns)	100	10	220	500	100	100	1
Edge time ratio K	100	10	220	500	100	100	1
$\triangle V_{OUT}$ (mV)	71	35	230	336	33	23.5	230
Min. PSR from DC to 50 kHz (dB)	−11	−22	−33	−22.5	−46	−70	−51.4
FoM (ps)	6.872	0.098	7.084	84.42	0.8118	0.705	0.0506

6. Conclusions

In this paper, an nA-level and 100 mV dropout LDO, combined with bulk modulation and adaptive power transistors is proposed and simulated in the 0.18 μm CMOS process. The operation principle, circuit implementation, stability analysis, and simulation results have been presented in detail. Particular attention has been paid to obtaining a low dropout and low supply voltage by the bulk modulation technology, without an auxiliary amplifier.

Micromachines **2023**, *14*, 998

Adaptive power transistors are proposed to switch under different load conditions to ensure circuit stability while reducing circuit power consumption. In addition, the adaptive bias mirroring the current from the amplifier itself with bounds is utilized to improve the transient response. Due to the superiority of ultra-low-power consumption, low supply voltage, and a fast response, the proposed OCL-LDO is suitable for use as a point-of-load regulator in energy harvesting.

Author Contributions: Validation, writing—original draft preparation: Y.Z. (Yuting Zhang); sotfware, visualization, data curation, conceptualization, methodology, writing—review and editing, project administration, funding acquisition, Y.Z. (Yanhan Zeng); data Curation; visualization, Q.G. All authors have read and agreed to the published version of the manuscript.

Funding: This work was supported in part by the Natural Science Foundation of Guangdong Province under Grant No. 2023A1515012900, in part by the Science and Technology Innovation Strategy Special Foundation of Guangdong Province under Grant No. pdjh2022a0402, in part by the Special Projects in Key Fields of Guangdong Education Department under Grant No. 2022ZDZX1019, and in part by the National Natural Science Foundation of China under Grant No. 62141414 and 61704037.

Data Availability Statement: Not applicable.

Conflicts of Interest: The authors declare no conflict of interest.

References

1. Bahramali, A.; Lopez-Vallejo, M. An RFID-Based Self-Biased 40 nm Low Power LDO Regulator for IoT Applications. *Micromachines* **2021**, *12*, 396. [CrossRef] [PubMed]
2. Wu, C.; Zhang, J.; Zhang, Y.; Zeng, Y. A 7.5-mV Input and 88%-Efficiency Single-Inductor Boost Converter with Self-Startup and MPPT for Thermoelectric Energy Harvesting. *Micromachines* **2023**, *14*, 60. [CrossRef] [PubMed]
3. He, Z.; Luo, P.; Wang, H.; Chen, J.; Song, H. A wide-input-range, low quiescent current LDO with ED reference for piezoelectric energy harvesting. *AEU-Int. J. Electron. Commun.* **2022**, *157*, 154419. [CrossRef]
4. Ruan, T.; Chew, Z.J.; Zhu, M. Energy-Aware Approaches for Energy Harvesting Powered Wireless Sensor Nodes. *IEEE Sens. J.* **2017**, *17*, 2165–2173. [CrossRef]
5. Paul, S.; Honkote, V.; Kim, R.G.; Majumder, T.; Aseron, P.A.; Grossnickle, V.; Sankman, R.; Mallik, D.; Wang, T.; Vangal, S.; et al. A Sub-cm3 Energy-Harvesting Stacked Wireless Sensor Node Featuring a Near-Threshold Voltage IA-32 Microcontroller in 14-nm Tri-Gate CMOS for Always-ON Always-Sensing Applications. *IEEE J.-Solid-State Circuits* **2017**, *52*, 961–971. [CrossRef]
6. Reyes, A.C.; Abuellil, A.; Estrada-López, J.J.; Carreon-Bautista, S.; Sánchez-Sinencio, E. Reconfigurable System for Electromagnetic Energy Harvesting with Inherent Activity Sensing Capabilities for Wearable Technology. *IEEE Trans. Circuits Syst. II Express Briefs* **2019**, *66*, 1302–1306. [CrossRef]
7. Zou, Y.; Yue, X.; Du, S. A Nanopower 95.6% Efficiency Voltage Regulator with Adaptive Supply-Switching for Energy Harvesting Applications. In Proceedings of the 2022 IEEE International Symposium on Circuits and Systems (ISCAS), Austin, TX, USA, 27 May–1 June 2022; pp. 3557–3561. [CrossRef]
8. Carreon-Bautista, S.; Huang, L.; Sanchez-Sinencio, E. An Autonomous Energy Harvesting Power Management Unit with Digital Regulation for IoT Applications. *IEEE J.-Solid-State Circuits* **2016**, *51*, 1457–1474. [CrossRef]
9. Park, S.-Y.; Cho, J.; Lee, K.; Yoon, E. A PWM Buck Converter With Load-Adaptive Power Transistor Scaling Scheme Using Analog-Digital Hybrid Control for High Energy Efficiency in Implantable Biomedical Systems. *IEEE Trans. Biomed. Circuits Syst.* **2015**, *9*, 885–895. [CrossRef]
10. Ni, S.; Chen, Z.; Hu, C.; Chen, H.; Wang, Q.; Li, X.; Song, S.; Song, Z. An Output-Capacitorless Low-Dropout Regulator with Slew-Rate Enhancement. *Micromachines* **2022**, *13*, 1594. [CrossRef]
11. Hu, C.; Chen, Z.; Ni, S.; Wang, Q.; Li, X.; Chen, H.; Song, Z. A Fully Integrated Low-Dropout Regulator with Improved Load Regulation and Transient Responses. *Micromachines* **2022**, *13*, 1668. [CrossRef]
12. Poongan, B.; Rajendran, J.; Yizhi, L.; Mariappan, S.; Parameswaran, P.; Kumar, N.; Othman, M.; Nathan, A. A 53-μA-Quiescent 400-mA Load Demultiplexer Based CMOS Multi-Voltage Domain Low Dropout Regulator for RF Energy Harvester. *Micromachines* **2023**, *14*, 379. [CrossRef] [PubMed]
13. Man, T.Y.; Mok, P.K.T.; Chan, M. A High Slew-Rate Push–Pull Output Amplifier for Low-Quiescent Current Low-Dropout Regulators With Transient Response Improvement. *IEEE Trans. Circuits Syst. II Express Briefs* **2007**, *54*, 755–759. [CrossRef]
14. Pereira-Rial, Ó.; López, P.; Carrillo, J.M.; Brea, V.M.; Cabello, D. An 11 mA Capacitor-Less LDO With 3.08 nA Quiescent Current and SSF-Based Adaptive Biasing. *IEEE Trans. Circuits Syst. II Express Briefs* **2022**, *69*, 844–848. 2021.3130674. [CrossRef]
15. Ma, X.; Lu, Y.; Li, Q. A Fully Integrated LDO with 50-mV Dropout for Power Efficiency Optimization. *IEEE Trans. Circuits Syst. II Express Briefs* **2020**, *67*, 725–729. [CrossRef]

16. Silverio, A.A. Forward Body Bias Technique for Low Voltage and Area Constrained LDO Design in Deep Submicron Technologies. In Proceedings of the 2021 28th IEEE International Conference on Electronics, Circuits, and Systems (ICECS), Dubai, United Arab Emirates, 28 November–1 December 2021; pp. 1–4. [CrossRef]

17. Ming, X.; Zhou, Z.; Zhang, B. A low-power ultra-fast capacitor-less LDO with advanced dynamic push-pull techniques. In Proceedings of the 2011 IEEE/IFIP 19th International Conference on VLSI and System-on-Chip, Hong Kong, 3–5 October 2011; pp. 54–59. [CrossRef]

18. Gupta, M.; Srivastava, R.; Singh, U. Low Voltage Floating Gate MOS Transistor Based Differential Voltage Squarer. *Int. Sch. Res. Not.* **2014**, *2014*, 357184. [CrossRef]

19. Musa, F.A.S.; Nurulain, D.; Ahmad, N.; Mohamad Isa, M.; Ramli, M. Implementation of floating gate MOSFET in inverter for threshold voltage tenability. *Eur. Phys. J. Conf.* **2017**, *162*, 01069. [CrossRef]

20. Keikhosravy, K.; Mirabbasi, S. A 0.13-μm CMOS Low-Power Capacitor-Less LDO Regulator Using Bulk-Modulation Technique. *IEEE Trans. Circuits Syst. Regul. Pap.* **2014**, *61*, 3105–3114. [CrossRef]

21. Nagateja, T.; Kumari, N.; Chen, K.-H.; Lin, Y.-H.; Lin, S.-R.; Tsai, T.-Y. A 8-ns Settling Time Fully Integrated LDO with Dynamic Biasing and Bulk Modulation Techniques in 40nm CMOS. In Proceedings of the 2020 IEEE International Symposium on Circuits and Systems (ISCAS), Seville, Spain, 10–21 October 2020; pp. 1–4. [CrossRef]

22. Koo, Y.S. A design of low-area low drop-out regulator using body bias technique. *Electron. Express Lett.* **2013**, *10*, 1–12. [CrossRef]

23. Wang, Y.; Shu, Z.; Zhang, Q.; Zhao, X.; Chen, S.; Tang, F.; Zheng, Y. A Low-Voltage and Power-Efficient Capless LDO Based on the Biaxially Driven Power Transistor Technique for Respiration Monitoring System. *IEEE Trans. Biomed. Circuits Syst.* **2022**, *16*, 1153–1165. [CrossRef]

24. Burington, R.S. *Handbook of Mathematical Tables and Formulas*; McGraw-Hill: New York, NY, USA, 1973.

25. Tsividis, Y.; McAndrew, C. *Operation and Modeling of the MOS Transistor*; Oxford University Press: New York, NY, USA, 1999.

26. Cheng, Y.; Hu, C. *MOSFET Modeling and BSIM3 User's Guide*; Springer: New York, NY, USA, 1999.

27. Zeng, Y.; Huang, Y.; Luo, Y.; Tan, H.-Z. An ultra-low-power CMOS voltage reference generator based on body bias technique. *Microelectron. J.* **2013**, *44*, 1145–1153. [CrossRef]

28. Adorni, N.; Stanzione, S.; Boni, A. A 10-mA LDO With 16-nA IQ and Operating From 800-mV Supply. *IEEE J. -Solid-State Circuits* **2020**, *55*, 404–413. [CrossRef]

29. Ria, A.; Catania, A.; Bruschi, P.; Piotto, M. A Low-Power CMOS Bandgap Voltage Reference for Supply Voltages Down to 0.5 V. *Electronics* **2021**, *10*, 1901. [CrossRef]

30. Bu, S.; Leung, K.N.; Lu, Y.; Guo, J.; Zheng, Y. A Fully Integrated Low-Dropout Regulator With Differentiator-Based Active Zero Compensation. *IEEE Trans. Circuits Syst. Regul. Pap.* **2018**, *65*, 3578–3591. [CrossRef]

31. Bu, S.; Guo, J.; Leung, K.N. A 200-ps-response-time output-capacitorless low-dropout regulator with unity-gain bandwidth >100 MHz in 130-nm CMOS. *IEEE Trans. Power Electron.* **2018**, *33*, 3232–3246. [CrossRef]

32. Li, G.; Qian, H.; Guo, J.; Mo, B.; Lu, Y.; Chen, D. Dual active-feedback frequency compensation for output-capacitorless LDO with transient and stability enhancement in 65-nm CMOS. *IEEE Trans. Power Electron.* **2020**, *35*, 415–429. [CrossRef]

33. Li, S.; Zhao, X.; Dong, L.; Yu, L.; Wang, Y. Design of a Capacitor-less Adaptively Biased Low Dropout Regulator Using Recycling Folded Cascode Amplifier. *AEU-Int. J. Electron. Commun.* **2021**, *135*, 153745. [CrossRef]

micromachines

Article

A 7.5-mV Input and 88%-Efficiency Single-Inductor Boost Converter with Self-Startup and MPPT for Thermoelectric Energy Harvesting

Chuting Wu, Jiabao Zhang, Yuting Zhang and Yanhan Zeng *

School of Electronics and Communication Engineering, Guangzhou University, Guangzhou 510006, China
* Correspondence: yanhanzeng@gzhu.edu.cn

Abstract: This paper presents a single-inductor boost converter for thermoelectric energy harvesting. A two-stages startup circuit with a three-phase operation is adopted to obtain self-startup with a single inductor. To extract the maximum energy, a coarse- and fine-tuning MPPT is proposed to adaptively and effectively track the internal source resistance. By designing a zero-current detector, the synchronization loss is reduced, which significantly improves the peak efficiency. The boost converter is implemented in a 0.18-μm standard CMOS process. Simulation results show that the converter self-starts the operation from a TEG voltage of 128 mV and achieves 88% peak efficiency, providing a maximum output power of 3.78 mW. The improved MPPT enables the converter to sustain the operation at an input voltage as low as 7.5 mV after self-startup.

Keywords: thermoelectric energy harvesting; maximum power point tracking; boost converter; self-startup

1. Introduction

Citation: Wu, C.; Zhang, J.; Zhang, Y.; Zeng, Y. A 7.5-mV Input and 88%-Efficiency Single-Inductor Boost Converter with Self-Startup and MPPT for Thermoelectric Energy Harvesting. *Micromachines* **2023**, *14*, 60. https://doi.org/10.3390/mi14010060

Academic Editors: Qiongfeng Shi and Huicong Liu

Received: 26 November 2022
Revised: 20 December 2022
Accepted: 21 December 2022
Published: 26 December 2022

Energy harvesting technology can be used in passing sensing devices in internet of things (IoTs) [1–3] due to the self-power [4,5], which can provide the supply voltage instead of the battery. Recently, there is a lot of research concentrating on harvesting piezoelectric energy [6], radio frequency (RF) energy [7], photovoltaic energy [8] and thermoelectric energy [9–11]. Among them, thermal energy can be widely used in wearable devices [12] because the thermoelectric generator (TEG) can convert the temperature difference between human body and the environment into a voltage. However, the output voltage from TEG is as low as tens of millivolts when the temperature difference is ultra-low. Thus a boost converter is needed in the thermal energy harvesting system.

In recent years, several boost dc–dc converters have been reported for low power operation. Ref. [13] proposed a converter with a zero current switching (ZCS), which operates at the input voltages ranging from 20 mV to 250 mV. However, there is no MPPT circuit to extract the maximum energy and the startup process is completed by an additional source. Meanwhile, a MPPT techinique for variation tolerance was proposed in [14] to improve the overall efficiency, but the converter utilizes a battery to startup. Boost converters such as [15,16] can lower the input voltage to tens of millivolts. However, these converters fail to self-start from a low TEG voltage, which is unable to realize automatic operation.

To achieve coldstart, ref. [17] uses an off-chip transformer but increases the costs and sizes. The transformer-reuse technique restricts the peak efficiency. A fully integrated startup approach is adopted in [18,19]. However, the peak efficiency is only 58% and 76%, respectively, due to the improper loss control. Ref. [20] achieves a high peak efficiency of 83%, but it only starts from a TEG voltage of 220 mV, which is not low enough. Therefore, it is challenging to achieve a lower voltage startup and high efficiency with only one inductor.

This paper proposes a 88%-efficiency single-inductor boost converter for low power thermoelectric energy harvesting. A two-stage startup circuit with three phase operation is proposed to assist in self-startup of 128 mV. An accurate MPPT is proposed to ensure the maximum power harvesting even in a low input voltage of 7.5 mV, which, together with the improved zero current detector (ZCD), also can improve the end-to-end efficiency within a wide load range.

This paper is organized as follows. The principle of the proposed converter is introduced in Section 2. In Section 3, the concrete circuit implementation is presented. Section 4 shows the verification results, and the conclusion is drawn in Section 5.

2. The Principle of Proposed Converter

As shown in Figure 1, a dual-path boost converter using a single inductor is proposed. It mainly consists of several parts including the coldstart, MPPT, ZCD, voltage detectors and regulator. The proposed converter starts its operation from only 128 mV by the two-stage startup circuit. The stage-I startup circuit consists of the ring oscillator, clock booster and charge pump. The large duty cycle circuit and 3-stage multiplier are utilized in the stage-II startup.

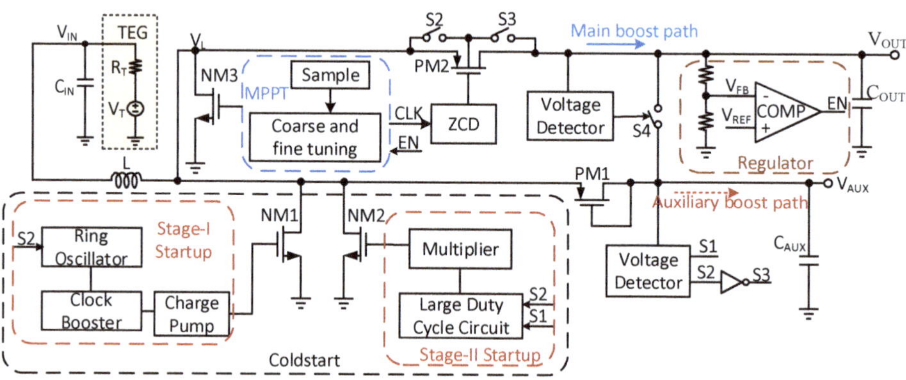

Figure 1. Architecture of the proposed converter.

2.1. Proposed Three Phase Operation

One of the challenges of this design is to support the low coldstart voltage while entering the normal operation. In previous cold-start works, the switched capacitor voltage multiplier, charge pump and voltage multiplier are usually utilized to improve the conversion efficiency for the initial boost, but their output power is low, which is not suitable for the main converter operation. However, none of the above converters meet the requirements of low input voltage, fast start boost and high conversion efficiency simultaneously. In this paper, a two-stage startup circuit with three phases is proposed to overcome the compromise between the low input voltage and efficiency for the TEG application. These output voltage levels are determined by the signal $S2$ and $S4$ from the voltage detectors. The illustration of the whole operation is presented in Figure 2.

In the startup phase, the stage-I startup self-starts from TEG supply and drives NM1, as shown in Figure 2a. Since the converter is weak, the load in this phase should be minimized. The ultra-low-power voltage detectors are designed, which is presented in Section 3.3.

In order to accumulate energy and generate a sufficiently high voltage at V_{AUX}, C_{AUX} is designed to be small enough (1 nF). Once V_{AUX} is higher than 330 mV, the stage-II startup is activated by S1 to drive NM2. The charging on C_{AUX} remains until V_{AUX} reaches 600 mV to trigger S2. The bulk of PM2 is biased to V_L during the self-starting, which avoids the insufficient current to the body diode. However, since V_{OUT} is not generated yet, a dual-path boost phase is needed to assist the converter to enter normal operation.

In the dual-path boost phase, the control modules are supplied by V_{AUX}. MPPT and ZCD are enabled by S2 to provide the clock for NM3 and PM2, as shown in Figure 2b. However, PM1 is a diode structure which causes a drop-off loss. Once V_{OUT} reaches 1 V, V_{AUX} and V_{OUT} are shortened by S4.

In the main boost phase, V_{OUT} supplies the control modules so that the converter can work independently, as shown in Figure 2c. Finally, the converter works in a steady state. NM3 and PM2 are temporarily turned off by EN once V_{OUT} exceeds 1.2 V. Both in the dual-path boost and main boost phase, the bulk of PM2 is biased to V_{OUT}, which can prevent the output current from reversing to the body diode. Figure 2d shows the waveforms of the whole operation.

The proposed dual-path converter adopts three phases to complete the coldstart and ultimately enters high-efficiency mode without an additional inductor.

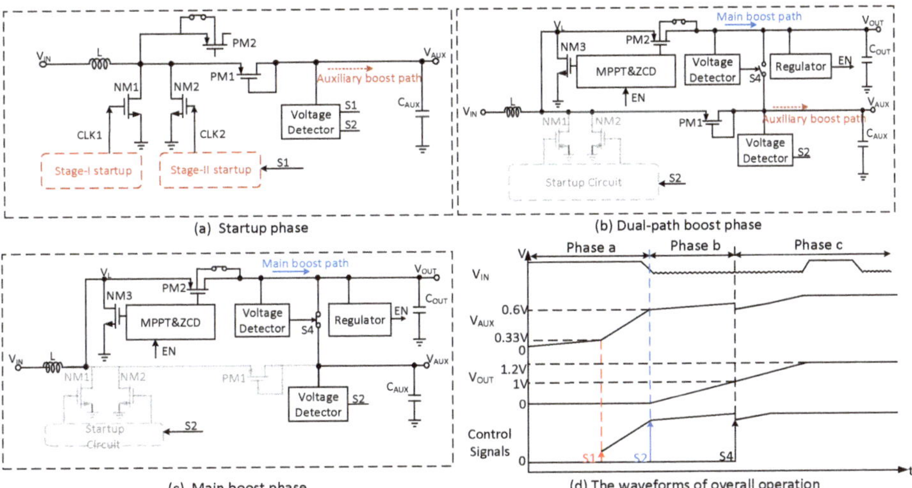

Figure 2. The three phases of the converter.

2.2. Efficiency Improvement

Another challenge is to improve the efficiency in normal operation. In this paper, the coldstart block is fully turned off to avoid the additional power consumption. To provide as much power as possible to the load with sufficiently high output voltage, the converter should first extract the maximum power from TEG, then efficiently transfer the power to the load. Owing to the input as low as tens of millivolts, the discontinuous conduction mode (DCM) is suitable for the converter to reduce switching loss.

2.2.1. Tracking Efficiency

To obtain the maximum input power, it is important to select the value of the inductor, the on-time of NM3, and the switching frequency. In DCM, the average inductor current can be expressed as

$$I_{AVE} = \frac{V_{IN}t_{LS}^2}{2L \times T},\tag{1}$$

where L is the value of inductor, t_{LS} is the on-time of NM3 and T is the working period. Therefore, the extracted input power is obtained as

$$P_{IN,ave} = V_{IN}I_{AVE} = \frac{V_{IN}^2 t_{LS}^2}{2L \times T}.\tag{2}$$

When the converter operates in the maximum power point (MPP), V_{IN} is equal to $V_T/2$. $P_{IN,ave}$ can be also expressed as

$$P_{IN,ave} = \frac{V_T^2 t_{LS}^2}{8L \times T} = \frac{V_T^2 t_{LS} \times D}{8L},$$

(3)

where D is the duty cycle of NM3, which is lower than 1. $P_{IN,MAX}$ is then given as

$$P_{IN,MAX} = \frac{V_T^2}{4R_T}.$$

(4)

Therefore, for a given MPP, the value of t_{LS} is depended on L. Large L requires large t_{LS}. However, large t_{LS} causes large ripple in V_{IN}, which decreases the tracking efficiency. Contrarily, small L needs low t_{LS}. The large duty cycle is required to transfer enough power to the load. Therefore, the switching frequency should be large, which increases the switching loss in the low input power. Considering the factors above, a 100 µH inductor is used to extract the maximum power and reduce power loss.

The fractional open circuit voltage (FOCV) method is adopted because of its convenience. In the MPPT mode, the internal resistor of the converter can be expressed as [16]

$$R_{IN} = \frac{2L}{t_{LS}^2 f}.$$

(5)

From this equation, t_{LS} or f can be adjusted to track MPP. Considering the efficiency in ultra-low input power, f is fixed and t_{LS} is adjusted in this paper.

As shown in Figure 3, t_{LS} is adjusted by the charging current. When R_{IN} is larger than R_T, V_{IN} is higher than $V_T/2$. Then t_{LS} raises to decrease R_{IN} for the impendence matching. Contrarily, t_{LS} can be reduced to increase R_{IN}. Because R_T depends on the temperature difference of TEG, the traceable R_{IN} should cover a wide range to keep MPP when R_T deviates from the design value by 12% [14]. To ensure a high tracking speed, coarse tuning is needed. However, coarse tuning leads to oscillation near MPP, which causes a large ripple in V_{IN}. Considering the boost ratio, R_{IN} can be also expressed as

$$R_{IN} = \frac{V_{IN}}{I_{L,ave}} = \frac{2L}{t_{LS}}\left(\frac{1}{1 - \frac{V_{IN}}{V_{OUT}}}\right).$$

(6)

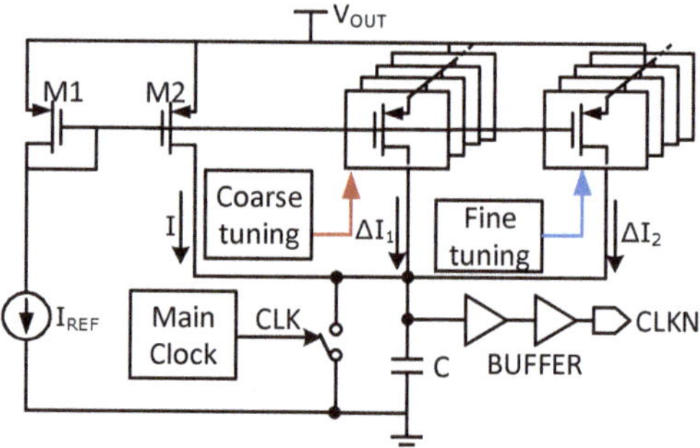

Figure 3. The three phases of the converter.

A small change of t_{LS} may decrease the boost ratio. To overcome these shortages, a fine-coarse tuning is proposed in this design, as shown in Figure 3. In this structure, t_{LS} can be expressed as

$$t_{LS} = \frac{CV_{th}}{I + \Delta I_1 + \Delta I_2},$$

(7)

where V_{th} is the threshold voltage of inverter. The range of R_{IN} is defined as R_{IN1}, which can be expressed as

$$R_{IN1} = \frac{2L(I + \Delta I_1 + \Delta I_2)^2}{(CV_{th})^2 f}, \tag{8}$$

where ΔI_1 is the current change in coarse tuning and ΔI_2 in fine tuning. To reduce the power loss in the open-circuit process, the sample period further decreases and less than the switching period. Since the input ripples may cause a deviation in MPP, which lowers the tracking efficiency, the input capacitor is added. The circuit implementation of MPPT is described in Section 3.2 in detail.

2.2.2. Conversion Efficiency

The conversion losses consist of conduction loss, switching loss, synchronization loss and power consumption of the control block [15]. When the input power is low, the switching loss is dominant, which can be expressed as

$$P_{sw} = (K_{LS}C_{LS} + K_{HS}C_{HS})fV_{OUT}^2, \tag{9}$$

where C_{LS} and C_{HS} are the gate capacitors of NM3 and PM2, respectively. K_{LS} and K_{HS} are the power consumption factor of the drivers, respectively, both of which are larger than 1. According to Equation (9), the switching loss remains unchanged when f is fixed in DCM.

When the input power is high, the dominant part is conduction loss. Since the on-time of PM2 can be neglected, the conduction loss can be expressed as

$$P_{cond} = f \int_0^{t_{LS}} I_L^2 R_{LS}\, dt, \tag{10}$$

where R_{LS} is the on-resistance of NM3 and I_L is the inductor current. When the converter works in the MPP mode, P_{cond} is derived as

$$P_{cond} = \frac{V_T^2 R_{LS} t_{LS}}{6LR_T}. \tag{11}$$

The conduction loss versus the maximum available power can be calculated:

$$\frac{P_{cond}}{P_{IN,MAX}} = \frac{2R_{LS}t_{LS}}{3L} \tag{12}$$

Equation (12) indicates that when R_{LS} is defined, the effect of P_{cond} on the conversion efficiency remains constant while V_T changes. To reduce R_{LS}, a large NM3 is used. To reduce the conflict between MPPT control and regulation, a burst control utilizing a dynamic comparator is adopted to reduce the conduction loss at light load.

Figure 4 shows the non-ideal delay of the PM2 switch. If the PM2 turns off early, a large overshoot will generate in V_L as shown in Figure 4a, which is a reduction in conversion efficiency. On the contrary, if PM2 turns off late, the output current will reverse to V_L as shown in Figure 4b, which also results in a large loss. Meanwhile, the ideal voltage conversion for the boost converter can be calculated as

$$V_{OUT} = V_{IN}\left(1 + \frac{t_{ON}}{t_{OFF}}\right) \tag{13}$$

It can be shown that for a high conversion ratio, t_{OFF} becomes very short, which causes a large loss when ZCD is not precisely designed.

To solve these issues, a static comparator with high speed and precision is adopted to control the turn-off time of PM2 dynamically. Meanwhile, the improper deadtime may cause a large shoot-through the current to damage the transistor, which must be controlled reasonably.

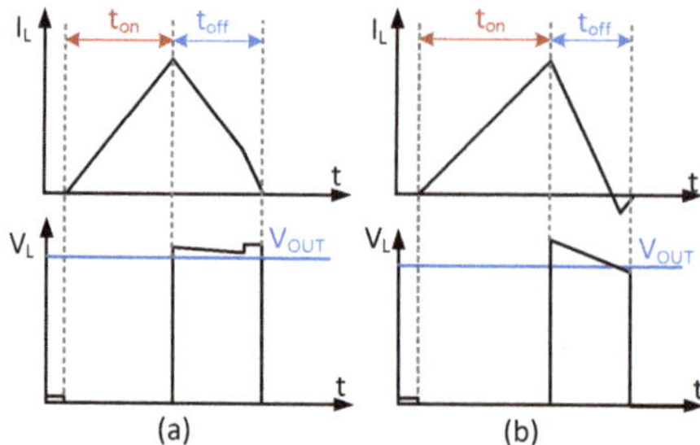

Figure 4. Detailed diagrams when (**a**) PM2 turns off early; (**b**) PM2 turns off late.

In this design, the following measures are taken to reduce the power consumption of control block. To ensure startup at a low input, the ultra-low-power voltage detectors are used. Meanwhile, because the synchronization loss leads to a low conversion efficiency, an accurate ZCD is needed in this paper, which consumes high power. This paper only enables ZCD within the off-time of NM3 to reduce the dynamic power consumption, which is described in Section 3.3.

3. Circuit Implementation

3.1. Low Voltage Startup

The coldstart is made up of the two-stages startup circuit. The proposed inductor-based stage-I startup circuit is depicted in Figure 5. It consists of ring oscillators, clock boosters and pelliconi charge pumps [21]. Due to the low supply voltage, the output power of the two charge pumps is limited in the pW level. Therefore, the clock used to drive NM1 is designed to be as low as 500 Hz. It is challenging for the system to produce a sustained clock for the charge pumps with a low supply voltage. A feasible scheme is the inverter-based oscillator. Figure 6 shows three kinds of inverters for the ring oscillators. The dc gain in the traditional inverter stage as shown in Figure 6a is given as

$$A_0 \geq \sqrt{1 + (\omega_1/\omega_0)^2}, \tag{14}$$

where ω_1 is the oscillation frequency and ω_0 is the 3 dB bandwidth of the inverter. A low supply voltage degrades the dc gain and the output swing of the inverter. It needs a lot of inverters, which results in low output frequency. The cascode inverter shown in Figure 6b and stacked inverter [22] depicted in Figure 6c can improve the dc gain and the output swing by dynamically reducing the leakage current in the discharging phase. The dc gain of the two inverters can be expressed:

$$A = \frac{(1 + \frac{g_{m1}}{g_{ds1}})g_{m2} + (1 + \frac{g_{m3}}{g_{ds3}})g_{m4}}{g_{ds2} + g_{ds4}} \tag{15}$$

where g_m and g_{ds} are the transconductance and channel conductance, respectively. The increased dc gain enhances the oscillation capability. However, as shown in Figure 6d, the frequency of the stacked-inverter-based oscillator is much smaller than another one, which lowers the output current of the charge pump. Therefore, this paper uses five cascode inverter stages to build the ring oscillator. The simulated frequency is 2.86 kHz at 128 mV supply.

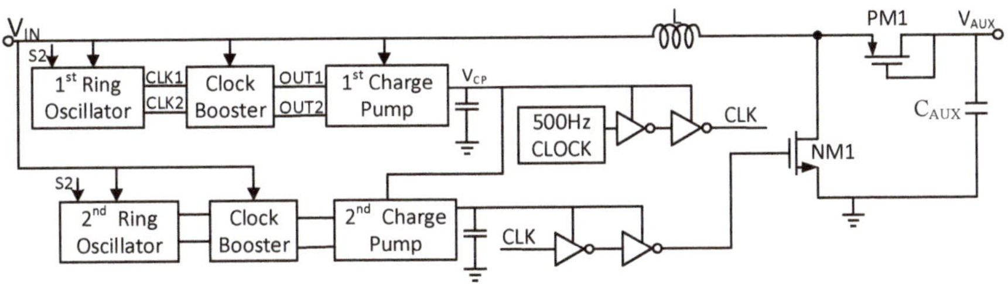

Figure 5. Stage-I startup circuit.

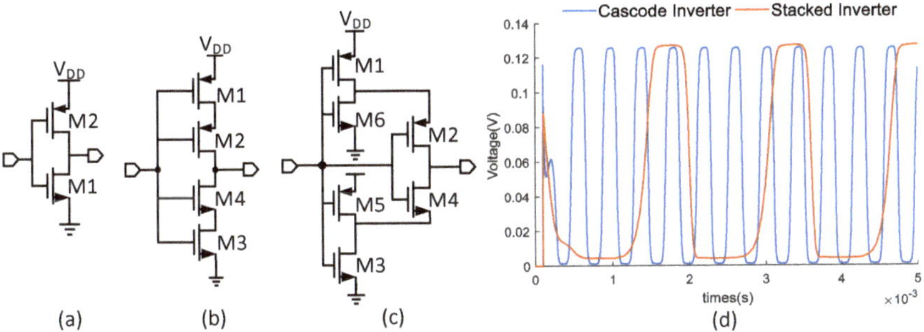

Figure 6. (**a**) Conventional inverter; (**b**) cascode inverter; (**c**) stacked inverter; (**d**) the oscillation of two ring oscillators.

As shown in Figure 7a, the clock booster with a higher amplitude is used to improve the charge–transfer ability. Once CLK2 is high, V_F and V_B step low while V_C and V_A are high. Hence, both C1 and C2 store a voltage of V_{IN}. Contrarily, C1 and C2 are connected in series, and V_D is boosted to $3V_{IN}$. Similarly, since CLK1 is opposite to CLK2, V_E can obtain $-V_{IN}$ when CLK2 is high. In this way, the clock amplitude is boosted to $4V_{IN}$. However, due to the loss and leakage current, the practical swing is smaller than $4V_{IN}$.

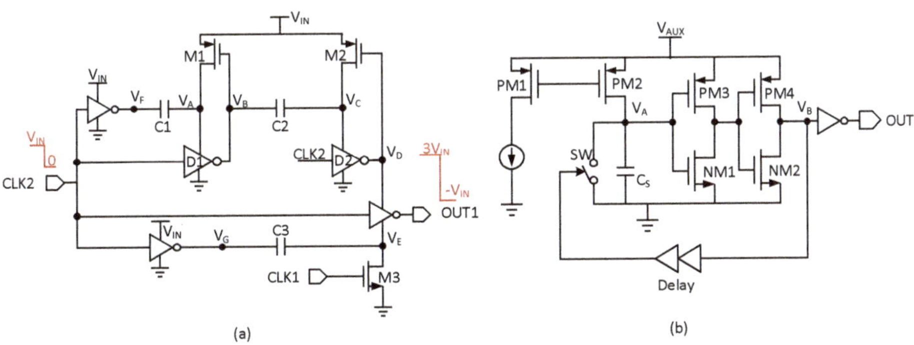

Figure 7. (**a**) The clock booster; (**b**) the large duty cycle circuit.

In stage-I startup, the two charge pumps are driven by the clocks OUT1 and OUT2 from the clock boosters, which is shown in Figure 8a. If the loss and load are zero, each charge pump could produce an output voltage at 9 times of V_{IN}. However, at a low supply, due to the poor clock driving capacity and the leakage current, the charge pump generates a lower boost ratio. Figure 8b shows the simulated results of stage-I startup. The 1st charge pump is driven to charge up V_{CP} to 450 mV with a 100 pA load. The boost ratio is limited to 3.5. Meanwhile, the boost ratio of the 2nd charge pump is limited to 1.5 with a 50 pA load. The efficiency of each charge pump is 4%. Finally, a clock with a 600 mV output swing is obtained.

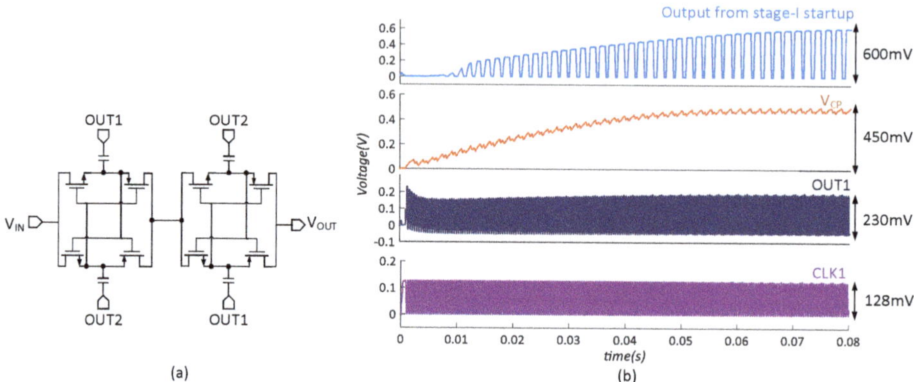

Figure 8. (**a**) The structure of each charge pump; (**b**) stage-I startup simulation waveform when $V_{IN} = 128$ mV.

The stage-II startup consists of the large duty cycle circuit and 3-stage multiplier, which is used to drive NM2. As shown in Figure 7b, when the large duty cycle circuit is activated, C_S is charged until V_A reaches the threshold voltage of the inverter. Then V_B goes high and turns on SW immediately to discharge C_S. Therefore, V_B is set to be a narrow pulse, and the output is a large duty cycle. Since the amplitude is not sufficient to drive NM2, the 3-stage multiplier is used to generate a 4-times output swing.

3.2. Accurate MPPT

Figure 9 shows the structure of the MPPT circuit. Considering the PVT variation, the duty of LS switch is adjusted by controlling the charging current. Once SW1 and SW2 are activated, V_T is sampled by disconnecting the converter from the input, and divided in half by the charge sharing of C_1 and C_2. The sampling frequency is set to be as low as $1/4096$ of the switching frequency. Therefore, the energy loss in the open-circuit process can be neglected.

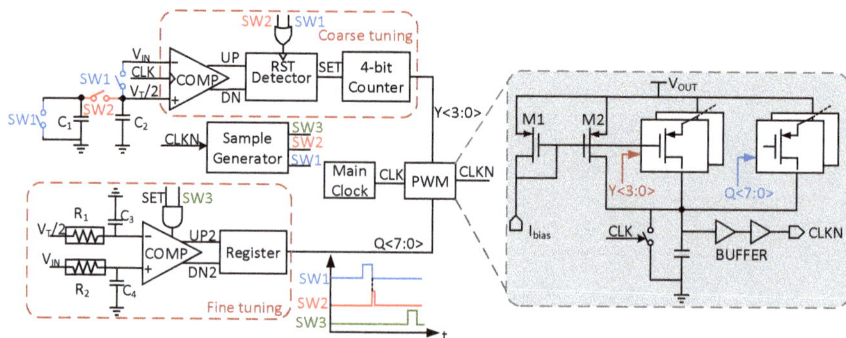

Figure 9. Schematic of the MPPT controller.

Then, V_{IN} is compared with $V_T/2$ to decide the duty. The coarse- and fine-tuning approach is adopted to obtain accurately and fast tracking. The dynamic comparator and 4-bit counter realize coarse tuning cycle by cycle, whereas the static comparator together with registers starts fine tuning once the coarse tuning is done. Therefore, the duty could be calibrated after several cycles to obtain the maximum energy. The ending signal of the coarse tuning is generated by the detector, which consists of two D flip-flops. When the UP or DN changes, the detector can sense this transition and set the signal SET to high.

A binary-weighted control with a step of 1 μs is employed in the coarse-tuning process. In fine-tuning, the static comparator controls two registers so that increments and decrements can be detected. One register increases the duty with Q<3:0>, while the other decreases the duty with Q<7:4>. Q<7:0> can control the duty cycle with a step of 200 ns.

The quiescent current of the static comparator is 200 nA. To further reduce the average power consumption, the static comparator only works when both SET and SW3 are simultaneously high. Since the internal resistor changes with the charging/discharging action, the ripples of V_{IN} exist, which lowers the accuracy of MPPT. Therefore, a low-pass filter is added to reduce this ripple in fine tuning. The off-chip resistors can be replaced by the pseudo-resistors, which is about 3.4 GΩ in 15-300 mV open-circuit voltage. Both C_3 and C_4 are 100 fF. Thus, the cutoff frequency is designed to be about 468 Hz. The simulated peak tracking efficiency is 99% with high accuracy and speed, consuming low power.

3.3. Zero Current Detector and Low Power Voltage Detector

The schematic of ZCD is shown in Figure 10a. It is composed of two one-shot circuits, a RS flip-flop and a static high-precision comparator from [23]. The voltages V_L and V_{OUT} are sensed by the comparator per cycle to detect the inductor current direction. Once the difference of two voltage-nodes decreases to 0, the HS switch is turned off immediately. According to Section 3, a proper dead time is needed and realized by the below one-shot circuit. For quickly and accurately tracking, the comparator consumes 6 μA quiescent current. Hence, the comparator is only enabled within the off-time of NM3 to reduce the average power consumption.

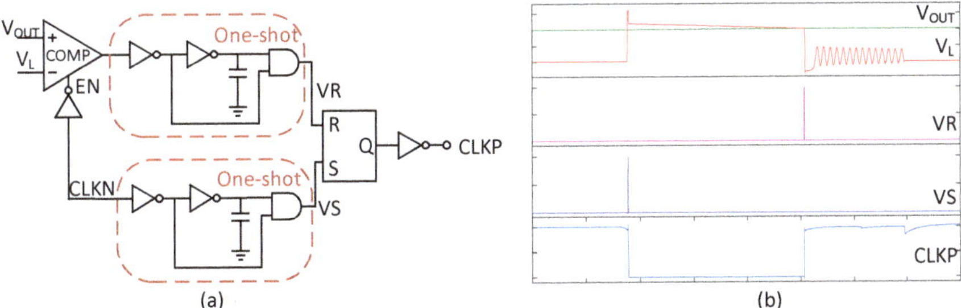

Figure 10. (a) The ZCD; (b) Simulated waveform of ZCD.

The voltage detector with its waveform is shown in Figure 11. When V_{DD} reaches $V_{Trigger}$, V_{OUT} turns high. The trigger voltage is determined by comparing I_{UP} and I_{DOWN}. To reduce the power consumption, M1-M4 are working in the subthreshold region. According to [24], I_{UP} is expressed as

$$I_{UP} = \mu C_{ox} \frac{W_1}{L_1} e^{\frac{V_{DD} - V_{TH1}}{m V_t}},\tag{16}$$

where μ is the hole mobility, C_{ox} is the gate oxide capacitance per area, m is a constant value of 1.1, V_t is approximately 26 mV at room temperature(T = 300 K), and V_{TH1} is the threshold voltage of M2–M4. I_{DOWN} can be expressed as

$$I_{DOWN} = \mu C_{ox} \frac{W_2}{L_2} e^{\frac{-V_{TH2}}{m V_t}},\tag{17}$$

where V_{TH2} is the threshold voltage of M1. Once I_{UP} is equal to I_{DOWN}, $V_{Trigger}$ turns high. If the threshold voltages of M1–M4 are equal and $V_{Trigger}$ is set to be 330 mV, the size ratio is derived as

$$\frac{W_2/L_2}{W_1/L_1} = 69. \tag{18}$$

However, if $V_{Trigger}$ increases to 1 V, the size ratio increases to 162,755, which is too large to implement. So a diode-connected PMOS can be added to M4, and its size is also W_1/L_1. Then under the same conditions, the size ratio decreases to 6250. As shown in Figure 11, the voltage detector can set the triggered voltage by adjusting the quantity of diode-connected PMOSs.

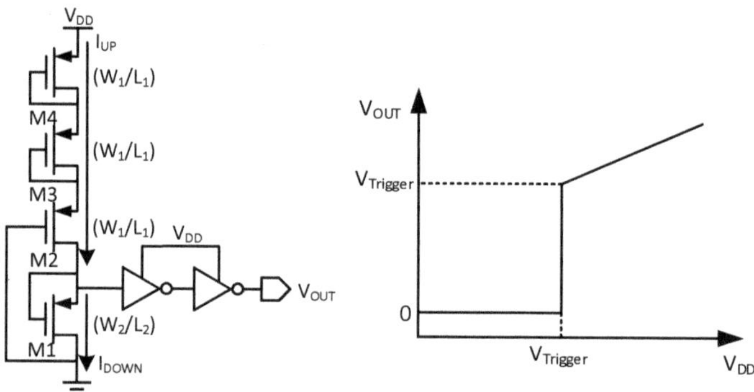

Figure 11. Schematic of the voltage detector and its waveform.

4. Circuit Verification

The proposed boost converter is implemented in a 0.18-μm standard CMOS process. The TEG is modeled as a DC power supply with an added series resistance of 5 Ω. A 100 μH inductor is used for the converter. To reduce the output ripples, the off-chip output capacitor is designed to be 10 μF.

Figure 12 shows the coldstart process, which demonstrates that the converter starts with a minimum open-circuit voltage V_T of 128 mV. Under this input condition, it takes 80.5 ms for V_{AUX} to reach 600 mV. After self-starting, the inductor current is shared by two boost paths. Once V_{AUX} ramps up to 1 V, V_{AUX} is shorted with V_{OUT} and the converter enters high-efficiency mode. Finally, the converter works temporarily and regulates V_{OUT} at 1.2 V.

Figure 13a shows the simulated results of the main clock and inductor current with 128 mV V_T. The frequency is 18.5 kHz and the on-time of LS is about 40 μs. Besides, the peak inductor current is 28 mA. The results show a negligible leakage in I_L, which indicates the proper current detection under a relatively heavy load condition with $I_{LOAD} = 600$ μA. Figure 13b shows the sample phase in 80 mV V_T with a 300 μA output current. Both LS and HS turn off, and thus, V_T can be sampled. The sample time is 1 ms and V_{OUT} decreases by 30 mV, which results in only 0.02% efficiency reduction in a sample period.

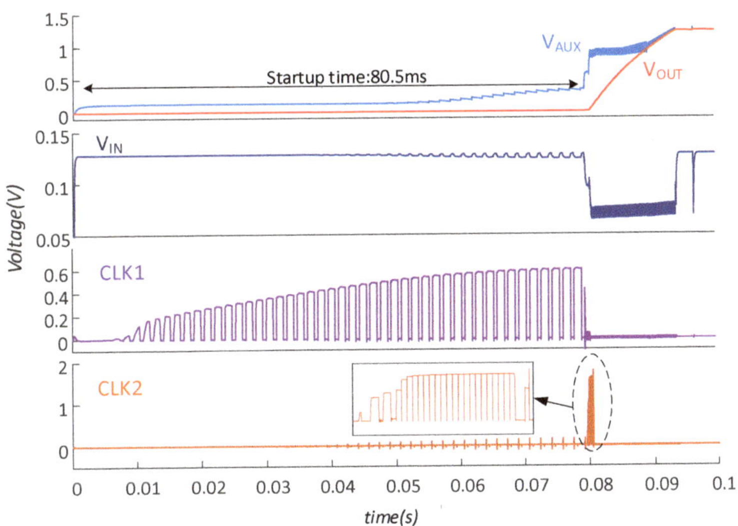

Figure 12. The waveform of startup process.

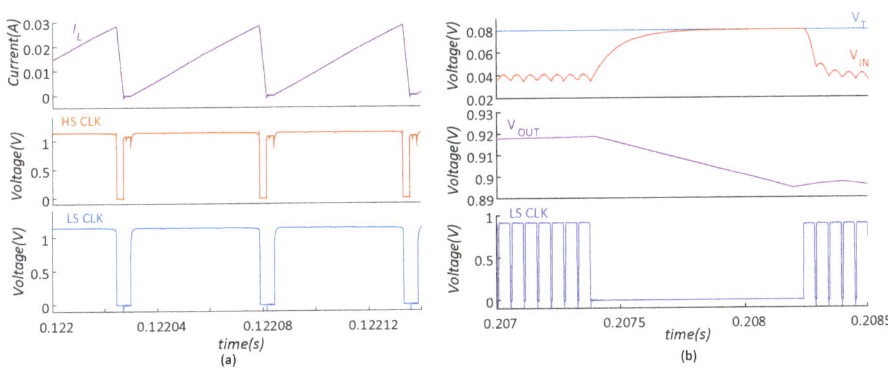

Figure 13. (**a**) The simulation results with V_T of 128 mV with 600 µA load; (**b**) the sample operation.

Figure 14 shows the system operation with MPPT when V_T changes from 300 mV to 15 mV. During 10~20 ms, the converter works intermittently at the open-circuit voltage of 300 mV, providing 2.5 mA output. After a 10 ms transition, the converter works with 15 mV V_T, providing 1.5 µA output. During 50~60 ms, V_T decreases to only 15 mV, while the minimum input voltage of proposed converter $V_{IN,MIN}$ is about 7.5 mV, which also means that the proposed MPPT still can efficiently work.

To verify the sensitivity of the proposed circuit to the process as well as temperature variations, and predict the performance after fabrication, the simulations with temperature ranging from 0 to 80 °C and different corners are carried out. Figure 15 shows the simulated output in FF corner at 0 °C and SS corner at 80 °C. The converter can normally work when V_T changes from 300 mV to 20 mV and finally provides a 1.08 V and 1.25 V regulated output voltage, respectively. In the tracking phase, the input voltage V_{IN} is about 10 mV, which indicates that the proposed MPPT can also efficiently and accurately work in different corners. Table 1 shows that the startup voltage increases to 130 mV in corner FF and 160 mV in the SS corner.

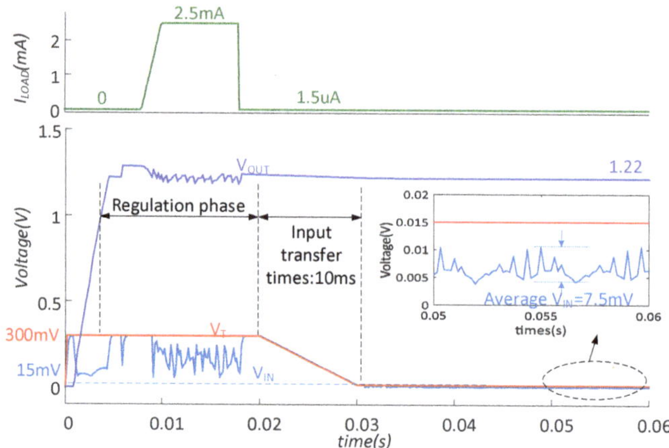

Figure 14. The simulated results with V_T from 300 mV to 15 mV.

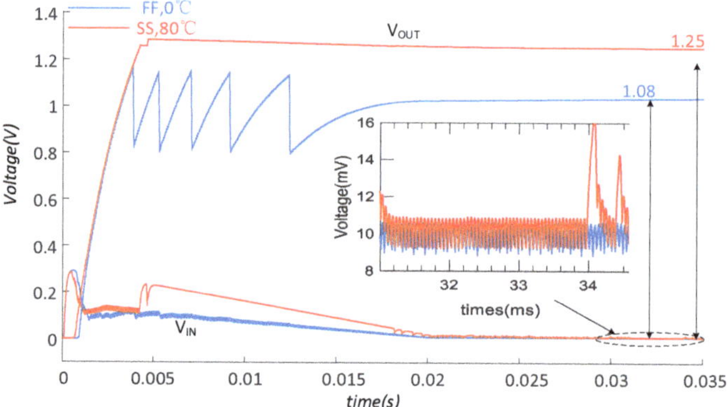

Figure 15. The simulation in different corners and temperature.

Table 1. Startup voltage in different corners.

Corner	TT	FF	SS
Startup voltage (mV)	128	130	160

Table 2 lists the tracking efficiency under the extreme temperatures and process corners. It can be observed that the tracking efficiency is greater than 93.6% when the input changes. Table 3 lists the conversion efficiency versus the light load current under different corners and temperatures. Meanwhile, it can be obtained from Figure 16a that the efficiency during the heavy load is not sensitive to the process. In the TT corner, the converter is able to sustain the operation at 15 mV V_T, which corresponds to only 1.8 µW output power. The efficiency is higher than 71.9% when V_T changes from 40 mV to 300 mV. In particular, the peak efficiency is 88% when V_T is 128 mV. Figure 16b shows the power consumption of each module when V_T is 128 mV. The ZCD and regulator account for the heaviest proportion of 76%. The major power consumption of the regulator is attributed to the two dividing resistors.

Table 2. Tracking efficiency under different corners and temperatures.

Corner	T	V_T	Tracking Efficiency
TT	0 °C	300 mV	99.7%
		15 mV	93.6%
	80 °C	300 mV	95%
		15 mV	98.9%
FF	0 °C	300 mV	99.1%
		20 mV	98%
	80 °C	300 mV	97.7%
		20 mV	99.8%
SS	0 °C	300 mV	99.1%
		20 mV	99%
	80 °C	300 mV	96.4%
		20 mV	99%

Table 3. Conversion efficiency versus light load current under different corners and temperatures.

V_T	I_{LOAD}	Corner	T	Efficiency
300 mV	150 µA	TT	27 °C	47%
		FF	80 °C	44%
		SS	0 °C	43%
150 mV	100 µA	TT	27 °C	46%
		FF	80 °C	36.5%
		SS	0 °C	50.1%
20 mV	2 µA	TT	27 °C	12.9%
		FF	80 °C	13.9%
		SS	0 °C	7%

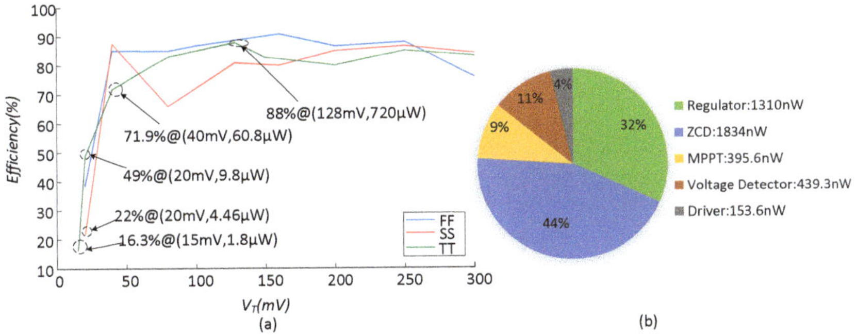

Figure 16. (**a**) The conversion efficiency and key output power in different corners in 27 °C; (**b**) the power consumption of each module.

Table 4 shows a performance comparison with the previous works. The proposed circuit provides a self-startup and regulated output with the maximum power of 3.78 mW, the minimum input voltage of 7.5 mV, and high efficiency of 88%.

Table 4. Performance comparison.

	This Work	**TCSII [25]**	**LSSC [26]**	**TCSII [27]**	**JSSC [15]**	**ISCAS [9]**
Year	2022	2019	2021	2017	2016	2019
Technology (nm)	180	180	130	65	180	250
Inductor (μH)	100	N.A.	N.A.	47	33	33
V_T (mV)	15–300	50–300	80–220	20–50	30–180	50–500
$V_{IN,MIN}$ (mV)	7.5	25	40	10	15	25
V_{OUT} (V)	1.2	1–1.6	0.3	≥ 1	1.9	3
R_T (Ω)	5	30	210	7	210	1
Peak $\eta_{Conv}@V_{IN}$	88% @64 mV	60% @60 mV	76.92% @80 mV	81% @25 mV	86.6% @40 mV	81% @150 mV
Self-startup voltage	128 mV	190 mV	300 mV	N.A.	N.A.	N.A.
Startup method	Coldtsart	Coldtsart	Coldtsart	External supplies	External precharge	External precharge
Maximum output power	3.78 mW	400 μW	N.A.	55 μW	30 μW	N.A.
MPPT	YES	YES	YES	NO	NO	YES
M or S result [1]	Simulation	Measurement	Measurement	Measurement	Measurement	Simulation

[1] Measurement or Simulation results.

5. Conclusions

A self-startup and single-inductor boost dc–dc converter for thermoelectric energy harvesting was presented and simulated in a standard 0.18 μm CMOS process in this paper. Based on the two stages with three-phase operation, the converter achieves self-startup at as low as 128 mV. Particular attention also was taken to obtain the accurate MPPT with as low as 7.5 mV input voltage by an adaptive coarse- and fine tuning. With the proposed MPPT and ZCD control, the converter has a peak efficiency of 88%, providing a maximum output power of 3.78 mW. These features together create the TEG-based harvester potential for self-powering wearable devices.

Author Contributions: Validation, writing—original draft preparation: C.W.; sofware, visualization: J.Z.; data curation: Y.Z. (Yuting Zhang); conceptualization, methodology, writing—review and editing, project administration, funding acquisition: Y.Z. (Yanhan Zeng). All authors have read and agreed to the published version of the manuscript.

Funding: This work was supported in part by the Special Projects in Key Fields of Guangdong Education Department under Grant No. 2022ZDZX1019, and in part by the National Natural Science Foundation of China under Grant No. 62141414.

Institutional Review Board Statement: Not applicable.

Informed Consent Statement: Not applicable.

Data Availability Statement: Not applicable.

Conflicts of Interest: The authors declare no conflict of interest.

References

1. Liu, H.; Zhang, J.; Shi, Q.; He, T.; Chen, T.; Sun, L.; Dziuban, J.A.; Lee, C. Development of a Thermoelectric and Electromagnetic Hybrid Energy Harvester from Water Flow in an Irrigation System. *Micromachines* **2018**, *9*, 395. [CrossRef] [PubMed]
2. Zulkepli, N.; Yunas, J.; Mohamed, M.A.; Hamzah, A.A. Review of Thermoelectric Generators at Low Operating Temperatures: Working Principles and Materials. *Micromachines* **2021**, *12*, 734. [CrossRef]

3.   He, W. A Shear-Mode Magnetoelectric Heterostructure with Enhanced Magnetoelectric Response for Stray Power-Frequency Magnetic Field Energy Harvesting. *Micromachines* **2022**, *13*, 1882. [CrossRef] [PubMed]
4.   Newell, D.; Duffy, M. Review of Power Conversion and Energy Management for Low-Power, Low-Voltage Energy Harvesting Powered Wireless Sensors. *IEEE Trans. Power Electron.* **2019**, *34*, 9794–9805. [CrossRef]
5.   Chong, Y.-W.; Ismail, W.; Ko, K.; Lee, C.-Y. Energy Harvesting for Wearable Devices: A Review. *IEEE Sens. J.* **2019**, *19*, 9047–9062. [CrossRef]
6.   Gao, S.; Huang, C.-Y.; Wu, L. Piezoelectric material based technique for concurrent force sensing and energy harvesting for interactive displays. *IEEE Sens.* **2017**, 1–3. [CrossRef]
7.   Liu, Z.; Hsu, Y.P.; Hella, M.M. A Thermal/RF Hybrid Energy Harvesting System with Rectifying-Combination and Improved Fractional-OCV MPPT Method. *IEEE Trans. Circuits Syst. I Regul. Pap.* **2020**, *67*, 3352–3363. [CrossRef]
8.   Chandrarathna, S.C.; Lee, J.W. A Self-Resonant Boost Converter for Photovoltaic Energy Harvesting with a Tracking Efficiency >90% Over an Ultra-Wide Source Range. *IEEE J. Solid-State Circuits* **2022**, *57*, 1865–1876. [CrossRef]
9.   Brogan, Q.; Ha, D.S. A Single Stage Boost Converter for Body Heat Energy Harvesting with Maximum Power Point Tracking and Output Voltage Regulation. In Proceedings of the 2019 IEEE International Symposium on Circuits and Systems (ISCAS), Sapporo, Japan, 26–29 May 2019; pp. 1–5. [CrossRef]
10.   Mu, J.; Liu, L. A 12 mV Input, 90.8% Peak Efficiency CRM Boost Converter with a Sub-Threshold Startup Voltage for TEG Energy Harvesting. *IEEE Trans. Circuits Syst. I Regul. Pap.* **2018**, *65*, 2631–2640. [CrossRef]
11.   Lim, B.M.; Seo, J.I.; Lee, S.G. A Colpitts Oscillator-Based Self-Starting Boost Converter for Thermoelectric Energy Harvesting with 40-mV Startup Voltage and 75% Maximum Efficiency. *IEEE J. Solid-State Circuits* **2018**, *53*, 3293–3302. [CrossRef]
12.   Bose, S.; Anand, T.; Johnston, M.L. A 3.5-mV Input Single-Inductor Self-Starting Boost Converter with Loss-Aware MPPT for Efficient Autonomous Body-Heat Energy Harvesting. *IEEE J. Solid-State Circuits* **2021**, *56*, 1837–1848. [CrossRef] [PubMed]
13.   Carlson, E.J.; Strunz, K.; Otis, B.P. A 20 mV Input Boost Converter with Efficient Digital Control for Thermoelectric Energy Harvesting. *IEEE J. Solid-State Circuits* **2010**, *45*, 741–750. [CrossRef]
14.   Kim, J.; Shim, M.; Jung, J.; Kim, H.; Kim, C. A DC-DC boost converter with variation tolerant MPPT technique and efficient ZCS circuit for thermoelectric energy harvesting applications. In Proceedings of the 2014 19th Asia and South Pacific Design Automation Conference (ASP-DAC), Singapore, 20–23 January 2014; pp. 35–36. [CrossRef]
15.   Katic, J.; Rodriguez, S.; Rusu, A. A Dual-Output Thermoelectric Energy Harvesting Interface with 86.6% Peak Efficiency at 30 μW and Total Control Power of 160 nW. *IEEE J. Solid-State Circuits* **2016**, *51*, 1928–1937. [CrossRef]
16.   Bandyopadhyay, S.; Chandrakasan, A. Platform architecture for solar, thermal, and vibration energy combining with MPPT and single inductor. *IEEE J. Solid-State Circuits* **2012**, *47*, 2199–2215. [CrossRef]
17.   Teh, Y.K.; Mok, P.K.T. Design of Transformer-Based Boost Converter for High Internal Resistance Energy Harvesting Sources with 21 mV Self-Startup Voltage and 74% Power Efficiency. *IEEE J. Solid-State Circuits* **2014**, *49*, 2694–2704. [CrossRef]
18.   Das, A.; Gao, Y.; Kim, T.T.H. A 220-mV Power-on-Reset Based Self-Starter with 2-nW Quiescent Power for Thermoelectric Energy Harvesting Systems. *IEEE Trans. Circuits Syst. I Regul. Pap.* **2017**, *64*, 217–226. [CrossRef]
19.   Goeppert, J.; Manoli, Y. Fully Integrated Startup at 70 mV of Boost Converters for Thermoelectric Energy Harvesting. *IEEE J. Solid-State Circuits* **2016**, *51*, 1716–1726. [CrossRef]
20.   Shrivastava, A.; Roberts, N.E.; Khan, O.U.; Wentzloff, D.D.; Calhoun, B.H. A 10 mV-Input Boost Converter with Inductor Peak Current Control and Zero Detection for Thermoelectric and Solar Energy Harvesting with 220 mV Cold-Start and −14.5 dBm, 915 MHz RF Kick-Start. *IEEE J. Solid-State Circuits* **2015**, *50*, 1820–1832. [CrossRef]
21.   Pelliconi, R.; Iezzi, D.; Baroni, A.; Pasotti, M.; Rolandi, P.L. Power efficient charge pump in deep submicron standard CMOS technology. *IEEE J. Solid-State Circuits* **2001**, *38*, 73–76. [CrossRef]
22.   Bose, S.; Anand, T.; Johnston, M.L. Integrated Cold Start of a Boost Converter at 57 mV Using Cross-Coupled Complementary Charge Pumps and Ultra-Low-Voltage Ring Oscillator. *IEEE J. Solid-State Circuits* **2019**, *54*, 2867–2878. [CrossRef]
23.   Zhang, X.; Zhang, Z.; Li, Y.; Liu, C.; Guo, Y.X.; Lian, Y. A 2.89 μW Dry-Electrode Enabled Clockless Wireless ECG SoC for Wearable Applications. *IEEE J. Solid-State Circuits* **2016**, *51*, 2287–2298. [CrossRef]
24.   Chen, P.H.; Ishida, K.; Ikeuchi, K.; Zhang, X.; Honda, K.; Okuma, Y.; Ryu, Y.; Takamiya, M.; Sakurai, T. Startup Techniques for 95 mV Step-Up Converter by Capacitor Pass-On Scheme and V_{TH}-Tuned Oscillator with Fixed Charge Programming. *IEEE J. Solid-State Circuits* **2012**, *47*, 1252–1260. [CrossRef]
25.   Chen, M.; Yu, H.; Wang, G.; Lian, Y. A Batteryless Single-Inductor Boost Converter with 190 mV Self-Startup Voltage for Thermal Energy Harvesting over a Wide Temperature Range. *IEEE Trans. Circuits Syst. II Express Briefs* **2019**, *66*, 889–893.[CrossRef]
26.   Wu, S.H.; Liu, X.; Wan, Q.; Kuai, Q.; Teh, Y.K.; Mok, P.K.T. A 0.3-V Ultralow-Supply-Voltage Boost Converter for Thermoelectric Energy Harvesting with Time-Domain-Based MPPT. *IEEE Solid-State Circuits Lett.* **2021**, *4*, 100–103. [CrossRef]
27.   Alhawari, M.; Mohammad, B.; Saleh, H.; Ismail, M. An Efficient Polarity Detection Technique for Thermoelectric Harvester in L-based Converters. *IEEE Trans. Circuits Syst. I Regul. Pap.* **2017**, *64*, 705–716. [CrossRef]

Review

A Review of Converter Circuits for Ambient Micro Energy Harvesting

Qian Lian [1,2], Peiqing Han [3] and Niansong Mei [1,2,*

[1] Shanghai Advanced Research Institute, Chinese Academy of Sciences, Shanghai 201210, China
[2] University of Chinese Academy of Sciences, Beijing 100049, China
[3] 3PEAK INCORPORATED, Shanghai 201203, China
* Correspondence: meins@sari.ac.cn; Tel.: +86-021-2032-5151

Abstract: The Internet of Things (IoT) has a great number of sensor nodes distributed in different environments, and the traditional approach uses batteries to power these nodes: however, the resultant huge cost of battery replacement means that the battery-powered approach is not the optimal solution. Micro energy harvesting offers the possibility of self-powered sensor nodes. This paper provides an overview of energy harvesting technology, and describes the methods for extracting energy from various sources, including photovoltaic, thermoelectric, piezoelectric, and RF; in addition, the characteristics of the four types of energy sources and the applicable circuit structures are summarized. This paper gives the pros and cons of the circuits, and future directions. The design challenges are the efficiency and size of the circuit. MPPT, as an important method of improving the system efficiency, is also highlighted and compared.

Keywords: energy harvesting; charge pump; rectifier; maximum power point tracking

Citation: Lian, Q.; Han, P.; Mei, N. A Review of Converter Circuits for Ambient Micro Energy Harvesting. *Micromachines* **2022**, *13*, 2222. https://doi.org/10.3390/mi13122222

Academic Editors: Qiongfeng Shi and Huicong Liu

Received: 5 November 2022
Accepted: 12 December 2022
Published: 14 December 2022

Publisher's Note: MDPI stays neutral with regard to jurisdictional claims in published maps and institutional affiliations.

1. Introduction

The demand for monitoring and management of smart homes, wearable devices, biomedical and industrial systems, and infrastructure has enabled the development of the Internet of Things (IoT) in recent years [1–3]. The IoT requires billions of wireless sensor network (WSN) nodes, which conventionally rely on battery power: however, the resultant problem is that the battery capacity is insufficient to last the entire sensor life cycle, and the labor and resources required to replace the widely distributed batteries require huge maintenance expense. In addition, the large size of the batteries, and environmental pollution, mean that this method of powering the IoT nodes is not the optimal solution. Consequently, technologies that harvest the ambient energy around sensors are gaining attention as power solutions for low-power IoT applications, with a view to replacing batteries for powering nodes.

The environmental energies that have been applied include solar, thermal, vibrational, and radio frequency (RF) energy, and they are obtained through photovoltaic cells (PVCs), thermoelectric generators (TEGs), piezoelectric harvesters (PEHs), and wireless energy harvesters (WEHs), respectively. The energy density of environmental energy sources, due to their fluctuating nature, is usually in the range of μW-mW/mm^2, with the possibility of going down to the pW order of magnitude under extreme conditions: for example, in an indoor environment, PV cells can harvest energy in the range of 10–100 μW/mm^2, while in an outdoor environment this value increases to 10–100 mW/mm^2. For piezoelectric and thermal energy, the environment in which the PEH and the TEG are located will also have an impact on the energy density harvested: if they are based on human harvesting, the resonant frequency of the PEH is lower, and the range of energy that can be harvested by the TEG is reduced, with the result that the energy density of both is in the μW order; however, if the PEH and the TEG are in an industrial environment—where the former has a higher resonant frequency, and the latter can harvest an increased temperature difference—then the energy density can reach the order of mW. Furthermore, RF energy is too weak, and it is difficult to harvest energy up to the mW level, but the advantage is that the energy is uninterrupted. Table 1 classifies the four types of energy harvesters according to their characteristics:

Table 1. Classifications of Energy Sources.

Type	PVC	TEG	PEH	WEH
Voltage	High	Low	High	High
Current	Low	High	Low	Low
Type	DC	DC	AC	AC
Drive Capability	General	Strong	General	Weak

DC, direct current; AC, alternating current.

The biggest problem with energy in the natural environment is its instability and weak nature. How to efficiently harvest environmental micro energy in a small area is the biggest challenge faced by the micro energy harvester system: for example, some energy sources have very low input voltage (<0.2 V), and conventional voltage converter systems suffer poor conversion efficiency or are even no longer valid; therefore, it is necessary to explore more effective boost circuits. This paper summarizes the respective characteristics and applicable topologies of the four energies, including the advantages and disadvantages of various boost converter structures and application scenarios—with special attention given to the applicability of different methods of maximum power extraction—and it also gives suggested future development directions. The rest of this article is organized as follows: Section 2 describes the circuit structures of the PV energy harvesting, and compares the pros and cons of the MPPT methods used in the micro energy harvesting scenario; Section 3 presents the differences between thermal energy harvesting and photovoltaic energy harvesting; in Section 4, the circuits applied to piezoelectric energy harvesting are introduced; Section 5 summarizes the structure of RF energy harvesting; in Section 6, the structures for simultaneous harvesting of multiple energies are described; Section 7 explains the directions and suggestions for future development; finally, Section 8 concludes the paper.

2. Photovoltaic Energy Harvesting

The benefit of solar energy is high density and ubiquity: the shortcoming is its nature, in that it is heavily influenced by the weather, and it is not readily available. Solar energy is generated via PVCs, to produce direct current, and a DC–DC converter transforms the unstable low voltage to a stable high voltage. Switching circuits, based on inductors or capacitors, can be used to implement DC–DC converters, with boost structures often used for inductor-based converters, and charge pump structures for capacitor-based converters. The capacitor-based converter is the main structure applied in solar energy collection systems, due to its ease of integration, low EMI, and flexible boost function. The output voltage of the converter can be regulated by a low dropout regulator. The unavailability of sunlight in dark environments makes the energy storage unit an integral part of the system, in order to maintain smooth operation. Assuming that the performance of the photovoltaic modules is in ideal conditions—such as adequate illumination and unshaded—then the following two factors will improve the efficiency of a solar energy harvesting system: (1) improving the efficiency of the DC–DC converter; (2) reducing the energy losses between the PV cells and the DC–DC converter, which can be achieved by the maximum power point tracking (MPPT) technique. However, when the power output is different between the PV cells, the generated power will be significantly restricted: this condition is known as "unbalanced generation". The imbalance may come from various sources, such as shadows from obstacles, moving clouds, and dust covering. Xiao et al. reported that this mismatch can have a disproportionate impact on system performance, because the module cell with the lowest output limits the current path through the other elements in the string, which can produce significant power losses. Consequently, for multi-source solar energy harvesting systems it is necessary to research anti-shading function, to cope with the losses due to the power imbalance between cells, arising from shading by trees or objects, that tends to occur in the application scenarios [4]. In high-power solar systems, this mismatch is usually solved with a flyback topology, whereas in microscale energy harvesting systems, this approach is too large and lossy.

To solve the problem of unbalanced solar power generation in micro energy harvesting systems, Jeong et al. have proposed an efficiency-aware collaborative multi-charging system that can improve the total conversion efficiency by adaptively distributing the input power among different cells, according to the power level, when the PV cells are partially shaded [5]. Kermadi et al. have proposed an MPPT technique whereby the MPP tracking is guaranteed under complex partial shading conditions [6].

For semiconductors, the types of loss include conduction and switching. Switching losses are positively correlated with frequency; however, the efficiency of MPPT decreases when the frequency is low; therefore, there is a frequency at which a balance between reducing switching losses and increasing the MPPT efficiency can be achieved. At the same time, the conduction losses are positively correlated with the forward conduction resistance. The three losses contribute differently to the total losses at different power levels: hence, the frequency can be adjusted appropriately at various system powers to achieve the lowest overall losses [7].

2.1. Structure of the Inductor-Based Energy Converter

Liu et al. have applied a dual-path six-switch (2P6S) converter with boost structure, as shown in Figure 1a. The 2P6S converter includes two paths: the direct path and the indirect path; the charge reaches the load from the PV cell through the direct path; the indirect path is from the PV cell through the battery to the load [8]. Wang et al. have proposed a dual-path three-switch (2P3S) converter based on the 2P6S structure, as shown in Figure 1b. The 2P3S converter reduces the number of charge-sharing power switches, but increases the direct path efficiency, while decreasing the indirect path power conversion efficiency (PCE); it also has the limitation that V_P and V_B must be greater than V_L [9]. To reduce the efficiency of the indirect path, Huang et al. have proposed a single-path three-switch (1P3S) converter, as shown in Figure 1c [10]. The disadvantages of the 1P3S converter are the lack of direct paths, and the constraint that V_P and V_L must be larger than V_B. The three configurations are suitable for different situations: 2P6S is the universal solution for most applications; 2P3S is the solution for a WSN with frequent operation; and 1P3S can be used in a WSN with a low duty cycle and standby power.

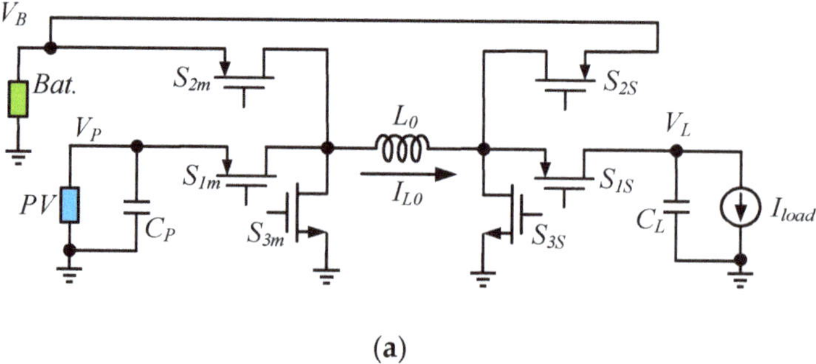

(a)

Figure 1. *Cont.*

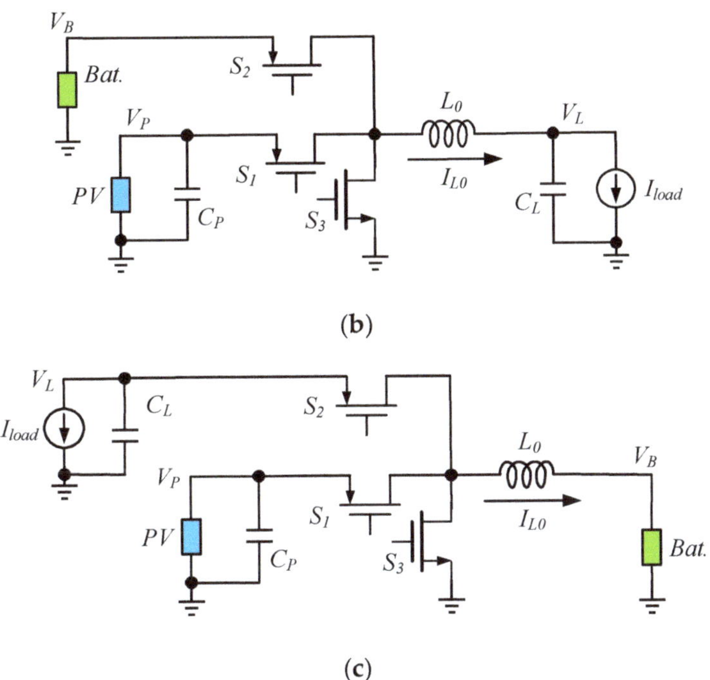

(b)

(c)

Figure 1. (**a**) 2P6S converter, (**b**) 2P3S converter, (**c**) 1P3S converter.

2.2. Charge Pump Topologies

Inductor-based converters are usually suitable for applications larger than 100 mW. The disadvantage of inductor-based technology is that it typically requires an off-chip inductor in the mH level, occupying a space of about 100 mm³, whereas a surface mount ceramic capacitor can take up a size of fewer than 0.5 mm³. In addition, capacitor-based structures have the benefit of being chosen for low-power-IoT and small-area applications, because they are easy to integrate. Given the cost efficiency of using capacitors, capacitor-based technologies have been developing in recent years.

The charge pump (CP) is one of the inductor-less DC–DC converters. This capacitor-based boost circuit was first proposed by Dickson in 1976, and is widely used in integrated circuits [11,12]. The chain of the Dickson CP is shown in Figure 2: each cell includes a charge transfer switch (CTS) and a pump capacitor, where the CTS acts as a switch, and consists of a diode or transistor [13].

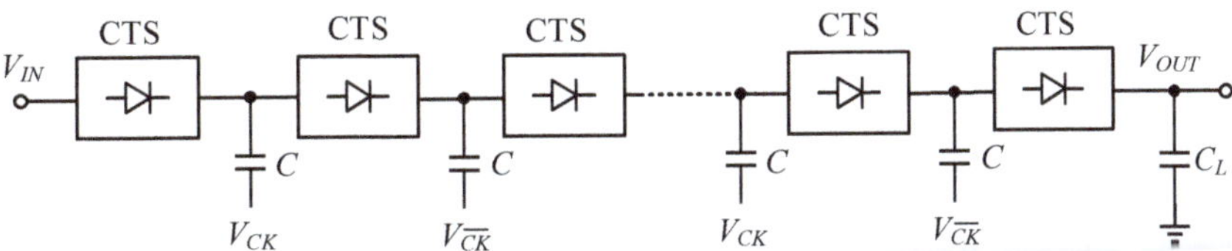

Figure 2. Simplified scheme of a linear Dickson charge pump.

The main factor that limits the efficiency of the Dickson CP structure and the output voltage is the threshold voltage (V_{TH}), which can be seen in Equation (1):

$$V_{OUT} = (V_{IN} - V_{TH}) + N\left(\frac{V_{CK}}{1+\alpha_T} - V_{TH}\right) - N\frac{I_{OUT}}{f_{CK}(1+\alpha_T)C'} \tag{1}$$

where V_{CK} is the voltage of the clock pulse, α_T is the ratio of the CTS parasitic capacitance to the pump capacitance, I_{OUT} is the current at the load, and f_{CK} is the clock frequency. To improve the CP's conversion efficiency, novel techniques are proposed.

2.3. Efficiency-Improving Topologies

2.3.1. Gate Biasing and Body Biasing Techniques

Controlled-switch technology using a CP was applied to memory for the first time. A simplified structure of the gate biasing technique is shown in Figure 3. In [14,15] it is used to generate positive and negative voltages efficiently: the difference is that NMOS is employed for positive generation, and PMOS is applied for negative generation. This scheme is characterized by the need to use a four-phase non-overlapping clock and an auxiliary boost capacitor. In this process, the gate voltage is constant, and is not affected by other voltage fluctuations, and M_i can conduct, as long as the gate-to-source voltage is higher than the V_{TH}.

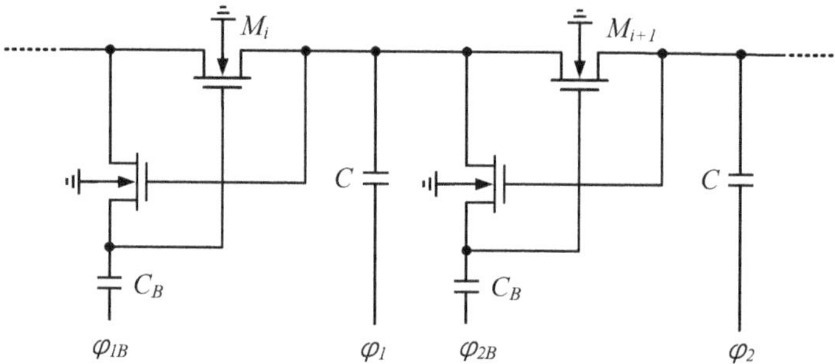

Figure 3. Gate biasing CP.

As the MOS switch in a bootstrap CP can replace the MOS diode in a Dickson CP, the ON resistance can be significantly reduced; however, it is difficult for the bootstrap CP to continue to achieve high conversion efficiency when the input voltage V_{IN} is below 100 mV. Fuketa et al. have improved the bootstrap CP, and have proposed a gate-boosted CP, where changing the signal amplitude of φ_{1B} and φ_{2B} from V_{IN} to the output voltage V_M can further reduce the ON resistance, and enhance performance [16].

The V_{TH} is affected by the difference between the substrate and the source voltage; therefore, the bias of the wells of the MOSFET can be employed to reduce the V_{TH} so that the effective input voltage of the CP is further reduced. The body biasing solution was first proposed by Sawada et al. in [17], as shown in Figure 4, where the source and the substrate of the MOSFET are shorted together. This connection allows the MOS to have the same threshold, and decreases the reverse losses.

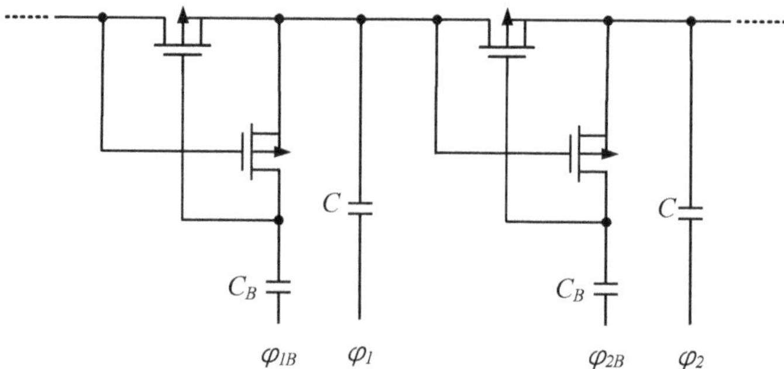

Figure 4. Body biasing CP.

2.3.2. Cross-Coupled Charge Pump

The cross-coupled CP, also called the latched CP, evolved from a two-stage CP, shown in Figure 5a, and is usually used as a voltage doubler or shifter: it was first proposed by Nakagome et al. as a word line driver with feedback, and so the structure is also known as Nakagome's cell [18]. Figure 5b shows the coupling of Nakagome's cell with a double series MOSFET, which was first published by Gariboldi and Pulvirenti as a monolithic quad line driver, a structure that constitutes a unit of a cross-coupled CP [19].

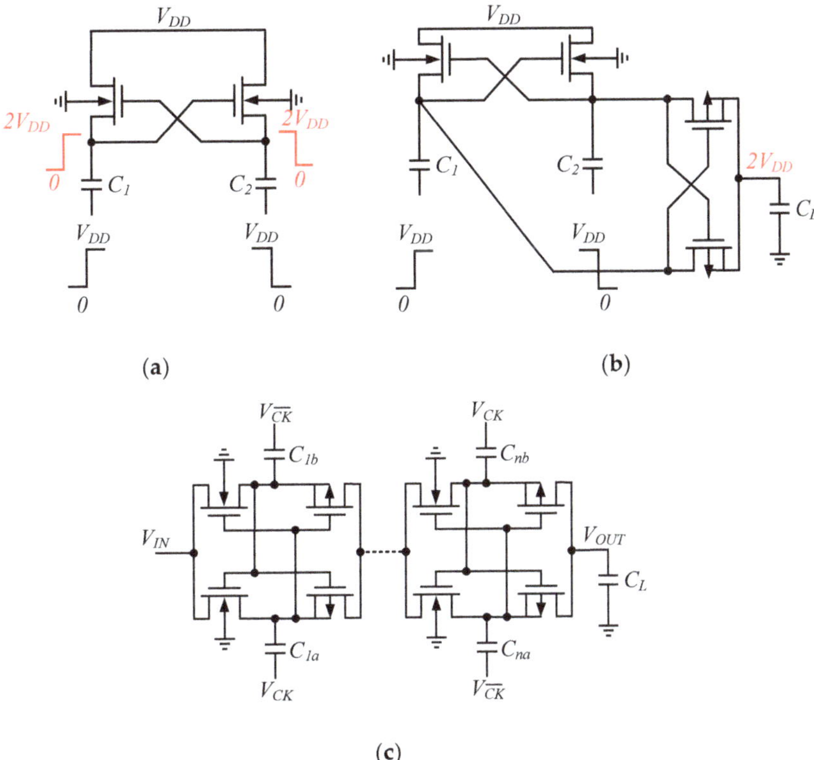

Figure 5. Nakagome's cell scheme: (**a**) without PMOS switches; (**b**) with PMOS switches; (**c**) a cross-coupled charge pump.

Cross-coupled CPs, as shown in Figure 5c, are similar in structure to dual-branch circuits; therefore, the pumping capacitance of a single-stage cross-coupled CP is half that of a traditional Dickson CP, so the corresponding MOSFET size is smaller. The cross-coupled structure reduces circuit ripple, and increases charge transfer efficiency. In addition, gate biasing and body biasing techniques can be superimposed on the cross-coupled CP structure, to further improve the PCE. The MOSFET is in a linear condition in the operating state, which turns off when in the inverted state.

Based on the above, many literatures have proposed modification of the CP structure. In [20], the charge current of a three-stage cross-coupled CP without and with a phase-shifted clock was derived, and the experimental results showed a threefold reduction in the charge current ripple, and achieved a PCE of 69%. Tsuji et al. proposed a low-leakage current driver matched with a three-stage CP to achieve high output voltage, wide load current, and 70.3% peak efficiency [21]. Peng et al. utilized both gate biasing and body biasing techniques to operate the CP circuit in the subthreshold region, with operating voltages as low as 320 mV in a 0.18 μm standard process [22]. In addition, forward body biasing (FBB) was first proposed by Chen et al. in [23], and this technique, when applied to a three-stage CP circuit, can achieve an output current improvement rate of 150% at an input voltage of 0.18 V. The dynamic body biasing (DBB) technique employed in [24] enhances the current transfer level, by preventing the reverse current of the cross-coupled NMOS, resulting in a 240% increase in the maximum output current, compared to using only conventional FBB.

2.3.3. Clock-Boosted Charge Pump

An additional improvement to the Dickson CP is the clock boost method. A quantitative analysis of the dynamic performance of the CP is presented in [25], which shows that decreasing the number of stages facilitates a reduction in rise time: however, this mechanism is at the expense of reducing the output voltage amplitude. The solution is to increase the clock voltage, which can reach twice the amplitude of the supply voltage by cascading Nakagome's cells, as in Figure 6b. A CP circuit with a clock booster (CKB) can reduce the rise time while maintaining the same area, or cut the area of the silicon used while keeping the rise time constant, and the improvement rate can reach 60%; however, the limited drive capability of the CKB potentially weakens the speed benefits of the reduced number of stages.

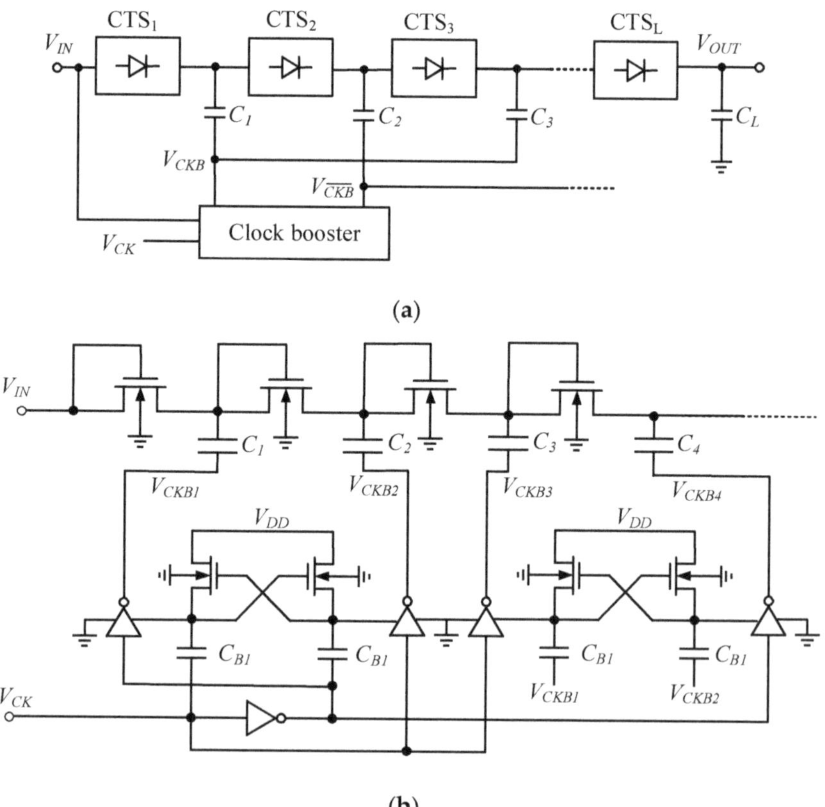

Figure 6. Clock-boosted CP: (**a**) simplified scheme; (**b**) multi-stage scheme.

Yi et al. proposed a differential bootstrapped ring-VCO (BTRO). The BTRO can harvest energy at low input voltages, to achieve a 1:10 boost ratio [26]. A BTRO composed of three inverter delay cells can generate a six-phase clock signal with swing boost, reducing the number of driven CP stages to only three, effectively increasing the PCE while significantly reducing the dependence of the clock frequency on the load capability. Each inverter delay unit that makes up the BTRO consists of two voltage triplers with two pairs of differential outputs to drive the delay unit and the CP element, respectively.

2.4. MPPT Techniques

The I/V characteristics of PVCs are nonlinear and have only one MPP, a value that varies with light level and temperature. To enable the circuit to continuously capture the MPP, and thus extract the maximum power, the MPPT technique was developed. The PV MPPT techniques are summarized in [27], including the hill climbing algorithm, the perturbation and observation (P&O) algorithm, conductance increment, fractional open-circuit voltage (FOCV), fractional short-circuit current (FSCC), and so on, to a total of 19 MPPT methods.

As the implementation of MPPT in micro energy harvesting systems needs to satisfy the characteristics of low cost, small size, and high efficiency, Han analyzed and compared the best fixed voltage (TBFV), FOCV, FSCC, load current maximization (LCM), and P&O methods that are applicable to micro systems. The comparative results show that the FOCV method is simple and inexpensive in circuitry, consumes less power, and can meet sufficient accuracy requirements, making it the most commonly used MPPT method in micro energy harvesting systems [28]. In addition, the P&O and hill climbing methods, as well as the artificial neural network (ANN) method, have been used extensively in solar energy harvesting.

2.4.1. Principle of the FOCV Method

The V_{MPP} of PV arrays are linearly related to V_{OC} at different irradiance and temperature conditions, which is the principle of the FOCV method:

$$V_{MPP} \approx \alpha V_{OC} \tag{2}$$

where α is a constant, determined by the characteristics of the PV array, and takes a value between 0.71 and 0.78. The empirical value of α for a particular PV array needs to be determined in advance under different conditions, when being applied. The V_{OC} is extracted by briefly disconnecting the power converter, and then passing it through the operator to reach the MPP: however, power losses are incurred in this process, and additional circuitry and memory devices—usually large capacitors—are required to maintain power to the load during the disconnection circuit. Various approaches have been proposed, to solve these problems. To reduce the losses in the disconnected circuit, Kobayashi et al. propose that the voltage generated by the PN junction diode can reach 75% of the V_{OC}, which can avoid arithmetic, and directly approximate the V_{MPP} [29]. A delay block is added in [30], to reduce the sampling time, which can decrease the losses, and a flash ADC is used in [31], to quantize the V_{OC} for the purpose of cancelling large capacitors.

Although Equation (2) is an asymptotic formula, it is adequate for approximation in cases where low precision is required: it does not demand complex controls and operations, and can make a good compromise between cost and performance.

2.4.2. The Hill Climbing and P&O Algorithms

The hill climbing and P&O algorithms are two methods of performing MPPT optimization to produce the same effect. The hill climbing method provides perturbation, by changing the duty cycle of the power converter, while the P&O algorithm offers perturbation by altering the operating voltage of the PV array; both algorithms yield the same result when the converter and the PV array are connected [27]. With the gradual increase in the application of capacitive DC–DC converters for energy harvesting, research on the hill climbing and P&O methods is also increasing. Unlike FOCV, the hill climbing and P&O methods are applied without disconnecting the circuit. The on-chip implementation of the MPPT control system with the P&O method retains high performance characteristics with low design complexity [32].

To match the impedance of the boost converter with the PV cell, the switching frequency, switching width, conversion ratio, and capacitance value of the converter can be adjusted separately or simultaneously: this is called the hill climbing method. When only one variable is changed, it is called one-dimension (1-D) MPPT, and is used in [33,34]; two-dimension (2-D) MPPT [33] changes two variables; three-dimension (3-D) MPPT is used in [35] to track the MPP of the PV cell by adjusting three variables. The advantage of the hill climbing method is high accuracy, and the disadvantages are a complex control circuit, high power consumption, and slow convergence; another problem is that the maximum power extraction cannot be effectively performed when the environment is changing rapidly and the PV module is shaded, as with the P&O method.

The power consumption of the FOCV, P&O, and hill climbing methods is presented in [36]: the data show that these three MPPT methods consume energy in the order of μW.

2.4.3. Other Promising MPPT Control Methods

The negative feedback control (NFC)-based low-overhead adaptive MPPT (AMPPT) method was first proposed by Lu et al. in [37], with the benefit that it has a simple control circuit and low power consumption compared to the hill climbing method: however, the disadvantage is that it requires the design of a dedicated voltage-controlled oscillator (VCO), so that it is not suitable for non-custom cases. In order to improve the adaptability of the circuit system to PV cells, and to make the system more suitable for industrialization, Wang et al. have proposed an improved AMPPT method aided by a neural network (NN) model [38]. Unlike the electrical equivalent model traditionally used to simulate PV cells, the neural network-assisted adaptive MPPT (NN-A-AMPPT) uses data to train the NN model to simulate PV cells, and proposes a reconfigurable VCO based on BJT.

The application of artificial intelligence (AI) algorithms, such as NN and optimization, as well as hybrid algorithms for solar energy harvesting, is described and compared in [39], which includes ANN and particle swarm optimization (PSO), for accurately tracking MPPs in any atmospheric conditions. Yap et al. evaluated the pros and cons, unresolved issues, and technical implementations of AI-based MPPT techniques [40]. For low-power systems, AI algorithms require large amounts of data, with slow convergence and high power consumption. Three NN algorithms corresponding to solar energy harvesting were analyzed and compared in [41]. Ahmed et al. compared the effectiveness of four traditional MPPT algorithms and ANN algorithms, and the results showed that the artificial intelligence network had the best MPP tracking [42]. Tabrizi et al. have proposed a digital MPPT algorithm that can achieve maximum power transfer by varying the number of CP stages under different load conditions when using a CP as a DC converter, and can avoid disconnecting the circuit from the source, suitable for ultra-low power consumption [43].

3. Thermal Energy Harvesting

Similar to solar energy, thermal energy is also in the form of DC, but the energy harvesting based on thermal energy is more sensitive to losses. In addition, of the four energy sources mentioned in this paper, the driving ability of the circuit based on thermal energy harvesting is the strongest. Unlike PV cells, the simple series configuration of thermoelectric modules can extract most of the available power, even in the presence of rather large mismatches among the modules. In addition, the MPPT method applied to solar energy is suited to thermal energy. FOCV is the most commonly used MPPT method. The results of four MPPT algorithms—P&O, Incremental Conductance (INC), Open Circuit Voltage (OCV), and Short Circuit Current (SCC)—applied to the same thermoelectric energy harvesting system were analyzed in [44], and show that the OCV algorithm has the fastest charging speed; however, unlike solar energy harvesting, α is taken as 0.5 in the FOCV based on thermal energy harvesting.

Thermal energy is characterized by low harvesting voltage: the output voltage of TEGs that harvest thermal energy is only in the range of 40 to 100 mV/K [45], so a boost converter with a low starting voltage is needed, and because of the low starting voltage, converters with self-starting capabilities are desired.

A fully built-in low-voltage start-up boost converter was proposed in [46], which can achieve a minimum start-up voltage of 82 mV. The input voltage range is 60–460 mV, and the output voltage range is 1–1.8 V. It has the static power of microwatt level, and a peak efficiency of 78.55%, by using a modified Schmitt trigger low-leakage logic gate. Radin et al. co-designed the oscillator and CP to provide self-starting and efficient conversion at ultra-low input voltages, achieving a minimum start-up voltage of 11 mV, and an operating voltage of 7.3 mV, with a peak efficiency of 85% at 140 mV input voltage, enabling the autonomous and efficient operation of the on-body device provided by a TEG at a temperature in the order of 1 °C gradient [47].

As thermal energy is more sensitive to losses, zero-current switching (ZCS) technology is commonly used in converters, to further reduce system energy losses and improve efficiency [7,30,47]. ZCS is a soft-switching technology used to ensure that the current waveforms do not overlap during switch-on and switch-off, thereby greatly reducing switching losses, as opposed to hard-switching technology, which has higher switching losses [48].

4. Piezoelectric Energy Harvesting

A complete PEH system can be divided into a vibration source and an interface circuit, in which the vibration source consists of a current source, I_p, an intrinsic capacitance, C_p, and a resistance, R, in parallel. R can be neglected sometimes, as the value of R is much higher than C_p at the resonant frequency. Ambient vibration sources typically have frequencies below 500 Hz: as a result, an interface circuit is needed to convert AC power to DC power, as required by the WSN. PEH interfaces can commonly be divided into three categories: (1) energy storage device-free; (2) inductor-based; (3) capacitor-based.

4.1. Energy Storage Device-Free Interface

An energy storage device-free interface comprises a rectifier with a simple structure that is easy to implement for full interface integration. The rectifier is dominated by a full-bridge rectifier (FBR), which sets a high V_{TH} for the generated energy extracted by the circuit. Figure 7 shows the FBR circuit used for piezoelectric energy harvesting, and its corresponding waveform, respectively. The shaded part on the right shows the losses caused by the voltage flip, from which it can be seen that it significantly limits the power efficiency, especially at low excitation levels.

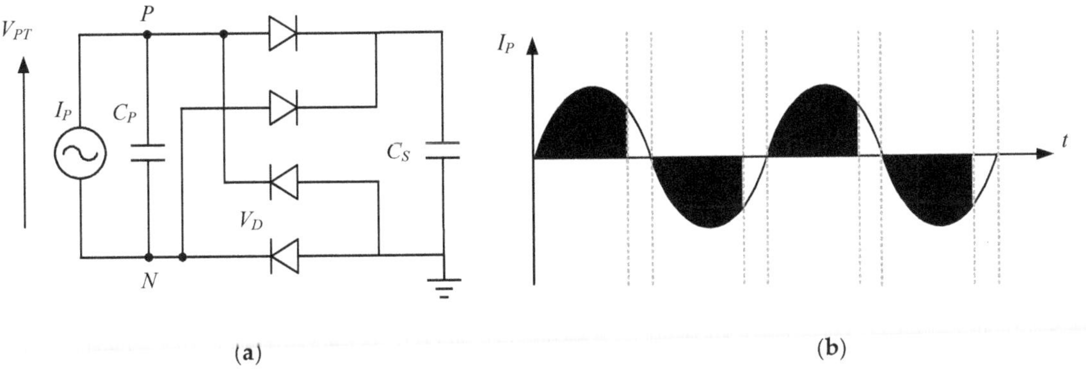

Figure 7. FBR (**a**) structure, (**b**) waveforms.

Ghovanloo et al. have proposed a fully integrated wideband high-current rectifier, replacing the two diodes in the FBR by two cross-coupled PMOS transistors. A comparison between the FBR and the PMOS rectifier with cross-coupling is shown in Figure 8 [49]. The gates of the cross-coupled PMOS transistors have a larger voltage swing drive than the diode-connected transistors, so a higher on/off current ratio can be achieved: however, the bottom two diodes are still implemented by the diode-connected transistors to block the reverse current, and the efficiency is not optimized [50].

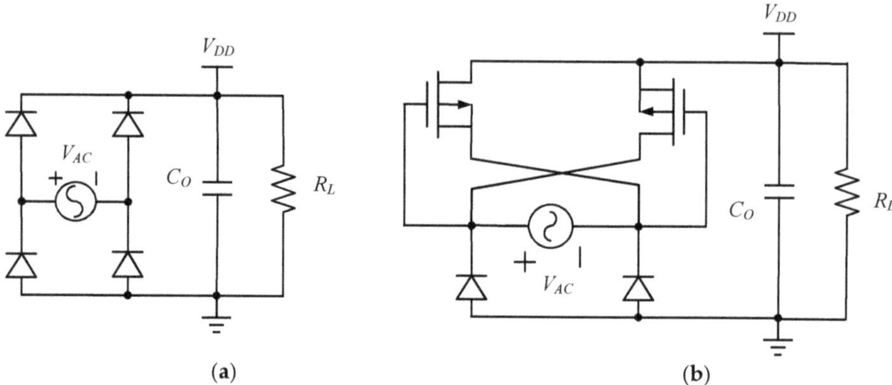

Figure 8. (a) FBR, (b) rectifier with cross-coupled PMOS.

To further reduce the reverse current, an active rectifier based on PMOS with cross-coupling is proposed, to improve the efficiency. The power losses due to the voltage drop of the forward voltage can be greatly reduced by replacing the diode with actively controlled transistors. Chang et al. have proposed a rectifier with active and non-overlapping control, as shown in Figure 9: the unbalanced comparator reduces the reverse current, and the non-overlapping control minimizes the additional losses caused by the transistors turning on and oscillating at the same time, and improves the power extraction efficiency and the voltage conversion ratio [51]. In this way, a PCE of 95% can be achieved at 20 kΩ and 200 kΩ load conditions.

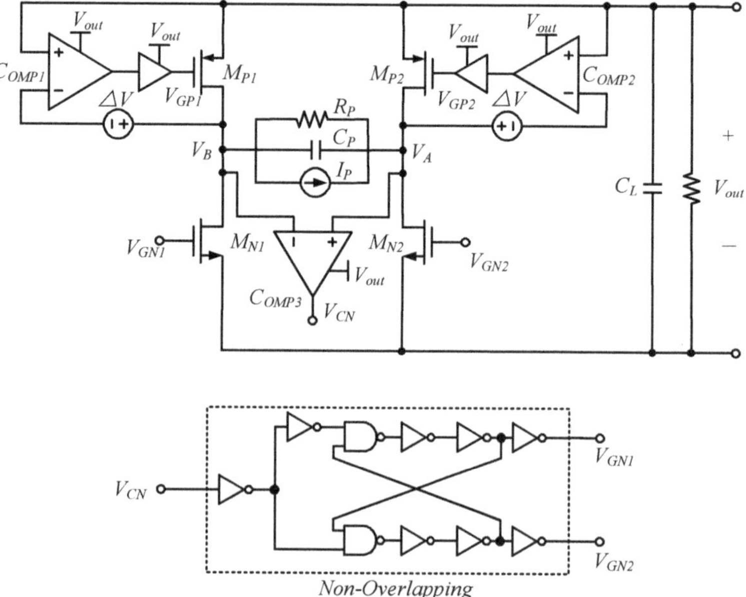

Figure 9. Active rectifier with non-overlapping control.

4.2. Inductor-Based Interface

Inductor-based rectification techniques include synchronous electrical charge extraction (SECE) [52], energy investing (EI) [53], and synchronous switch harvesting on inductor (SSHI) [54]. In order to improve the power extraction efficiency and reduce the losses due to the FBR, a phase-controlled switch and an inductor can be added to the FBR to form an SSHI circuit, with the degree of performance improvement defined by the maximum output improvement rate (MOPIR):

$$MOPIR = \frac{P_{rect,max}}{P_{FBR,max}} \tag{3}$$

where $P_{rect,max}$ and $P_{FBR,max}$ are the maximum output powers of the corresponding interface circuits at the rectifier and FBR outputs, respectively.

Figure 10 shows the structure of the SSHI circuit and its waveform: the synchronous pulse signal ϕ_{SSHI} controls L, and thus the RLC oscillation loop, to flip the voltage. The value of V_{PT} is $V_S + 2V_D$ or—$(V_S + 2V_D)$ before I_P crosses the zero point, and the voltage flips to $-V_F$ or V_F after the zero point. The absolute value of the resulting flip voltage, V_F, is always less than $V_S + 2V_D$, due to the resistive damping. V_F can be expressed as $V_F = (V_S + 2V_D)\exp(-\pi/\sqrt{(4L/(R^2C_P))-1})$ [55]. After the voltage flip, the value of $|V_F|$ continues to rise to $V_S + 2V_D$, during which the energy losses of the shaded part are generated. It can be seen that the losses generated by the SSHI circuit are greatly reduced, compared to the FBR. The efficiency of the SSHI circuit can be presented as:

$$\eta_{SSHI} = \frac{V_F}{V_S+2V_D} = exp\left(-\frac{\pi}{\sqrt{\frac{4L}{R^2C_P}-1}}\right) \tag{4}$$

where C_P is the intrinsic capacitance of the piezoelectric transducer (PT), L is the inductor, and R is the total resistance in the RLC loop, including the DC resistance of the inductor, the on-resistance of the switch, and other parasitic components.

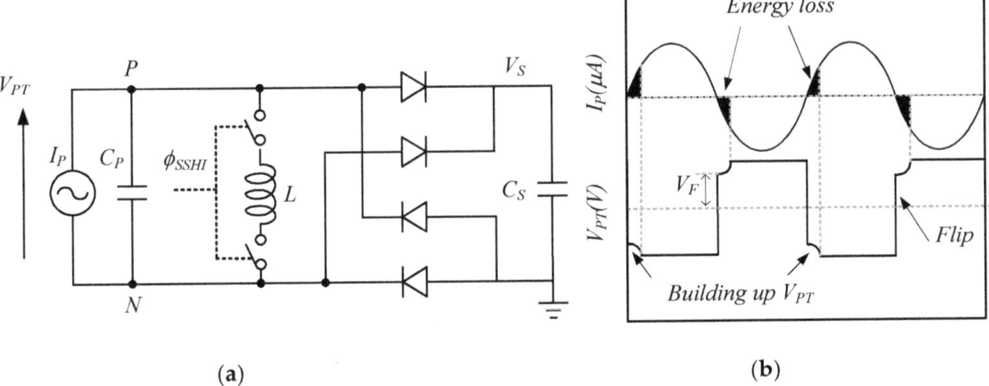

(a)   (b)

Figure 10. SSHI (a) structure, (b) waveforms [55].

Improving the performance of the SSHI circuit can be achieved by increasing the value of L and decreasing the value of C_P and R. It is worth noting that increasing L will result in a certain growth in R. The design of the SSHI in [54] can improve the energy extraction performance by an MOPIR of 5.8, with a DC output power of 40.6 μW and a power density of 8.12 mW/cm³.

SSHIs can be divided into series synchronous switch harvesting on inductors (S-SSHIs), and parallel synchronous switch harvesting on inductors (P-SSHIs). The SSHI synchronously flips the voltage on the PT, to minimize energy wastage due to internal capacitor charging and discharging, making it one of the most energy-efficient circuits, which ideally has no charge wastage; however, the efficiency of the SSHI does not reach 100%, due to the synchronous switching damping effect. Dell'Anna et al. have summarized the characteristics of other inductor-based rectifier schemes, such as SECE, Hybrid SSHI, and EI [56].

4.3. Capacitor-Based Interface

Capacitor-based circuits applied to piezoelectric energy harvesting include the synchronized switch harvesting on capacitor (SSHC) [57], the flipping-capacitor rectifier (FCR) [58], and the split-phase flipping-capacitor rectifier (SPFCR) [59], which allow on-chip integration by avoiding the use of off-chip inductors.

Figure 11 shows the circuit diagram of the SSHC interface with one charge-swap capacitor and its corresponding waveform. A switched-capacitor (SC) interface circuit uses three pulsed signals (ϕ_0, ϕ_n, and ϕ_P) to control five switches without overlapping. When the current crosses zero, and V_{PT} needs to be flipped from the high voltage $V_S + 2V_D$, then ϕ_P, ϕ_0, and ϕ_n conduct in turn, to bring V_{PT} down to $-1/3 \, (V_S + 2V_D)$ and vice versa, with a voltage flipping efficiency of $1/3$. The voltage flipping efficiency further increases when more SC modules are added. Du et al. has summarized the voltage flipping efficiency for an SSHC with 1–8 SCs, and the inductance required for an SSHI to achieve the same efficiency, respectively, which shows that an SSHC can significantly reduce the system size by using capacitors [55].

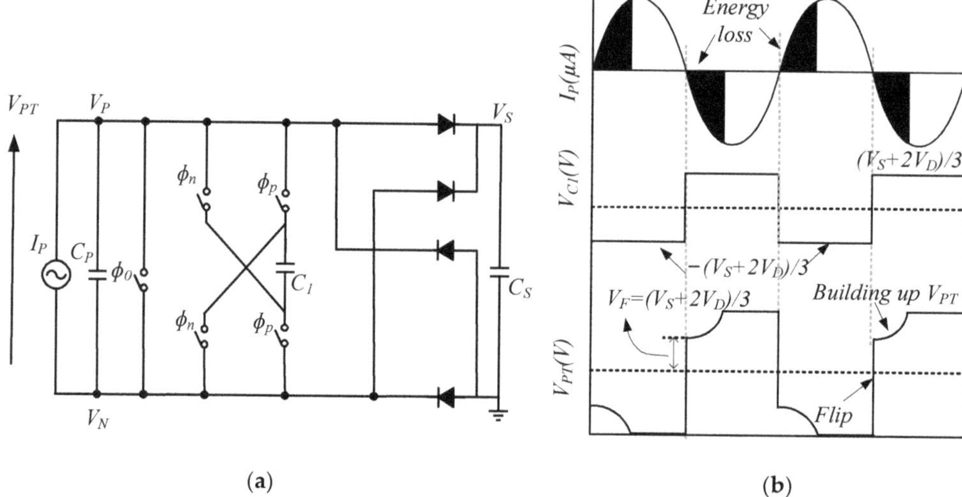

Figure 11. SSHC (**a**) with one charge-swap capacitor, (**b**) waveforms [55].

Capacitor rectification needs to achieve more flipping phases with less capacitors, to make the energy extraction efficiency as high as possible; however, too many flipping phases increase the control losses; therefore, the flipping phases need to be minimized without sacrificing the energy extraction efficiency. The SPFCR is proposed in [59], to achieve 21 flipping phases with four capacitors and MOPIR up to 9.3 times.

4.4. Maximum Power Extraction

Maximum power transfer occurs when the input impedance of the AC–DC converter is a complex conjugate of the PT source impedance; however, matching the capacitive component of the PT source with an inductor in the converter is impractical, as the inductance can be very large—in the tens to hundreds of the Henry range [60]. Li et al. proposed an up-conversion technique, to reduce the system area by impedance matching, after shifting the frequency of the energy generated by the PT source to a higher value [61]. On the other hand, because the source impedance varies with the excitation frequency, the load needs to operate at optimal conditions, in order to improve efficiency. This means that the DC–DC converter also needs to convert the rectified voltage according to the load requirements.

Ottman et al. used a DC–DC converter followed by the rectifier to regulate the load, but the circuit used two stages [62]. John Turner et al. used a bidirectional DC–DC converter with feedforward control to simulate a parallel RL load, but the circuit power dissipation was relatively large for small-scale systems [63]. Chew et al. implemented simultaneous MPPT and voltage regulation, using a single DC–DC converter [64]. The SSHI technique eliminates the capacitance term by voltage flipping, but does not reach the maximum power transfer. SSHIs and SSHCs can harvest energy efficiently from broadband excitation: however, the final power extracted is all load-dependent; hence, the variation of the load affects the efficiency.

A series of MPPT methods have been proposed. In [65,66] an FBR combined with the FOCV method, to extract the maximum power from the PT, was applied. To improve the rectifier efficiency, Kawai et al. designed a P-SSHI circuit with FOCV [67]. The FOCV method requires disconnecting the circuit to detect the open-circuit voltage: to reduce the power losses, Fang et al. proposed a method called fractional normal operating voltage (FNOV), based on S-SSHI, to avoid disconnecting the PEH from the load [68]. P-SSHI circuits combined with P&O MPPT techniques were proposed by Li et al. in [69], and a self-adjusting phase-shift-based SECE technique was proposed by Morel et al. in [70]. Wang et al. proposed an emerging MPPT technique based on envelope extraction without programmable control units and without disconnecting the circuit, with rectifiers using SSHI architecture [71]. An adjustable-delay SSHC method was proposed by Liao Wu et al. in [72], to regulate the power flowing into the load, by adjusting the pulse width of the switching pulse signal in the SSHC, avoiding the use of a second-stage circuit and inductors, and thus reducing the complexity of the circuit. In [59], a switched-capacitor DC–DC converter was composed of capacitors in the non-flipping state for load regulation, and a multi-voltage conversion ratio (MVCR) was obtained. An active rectifier circuit with two-dimensional P&O control was proposed in [73], which could track the maximum power point quickly, and be error-free.

5. RF Energy Harvesting

Compared to piezoelectric, RF energy has lower power density, yet the advantage is that it is not affected by energy irregularities, and has high reliability. The RF energy harvesting (RFEH) system can be classified into near-field RFEH and far-field RFEH: when the distance between transmitter and receiver is within the Fraunhofer distance, it is called near-field RFEH, and vice versa for far-field RFEH. Far-field RF is also known as ambient RF. The power density of far-field RF is lower than that of near-field RF.

Unlike near-field RF, far-field RFEH in the electric field (E) and magnetic field (H) has equal amplitude in the same point: hence, the received power of the antenna is quantifiable and predictable, and the received power can be calculated by

$$P_{RX} = \frac{P_{TX} G_{TX} G_{RX} \lambda^2}{(4\pi R)^2} \tag{5}$$

where P_{TX} is the transmitted power, G_{TX} is the transmitting antenna gain, G_{RX} is the receiving antenna gain, and R is the distance between the transmitting and receiving antenna. The architecture of the RFEH system is shown in Figure 12.

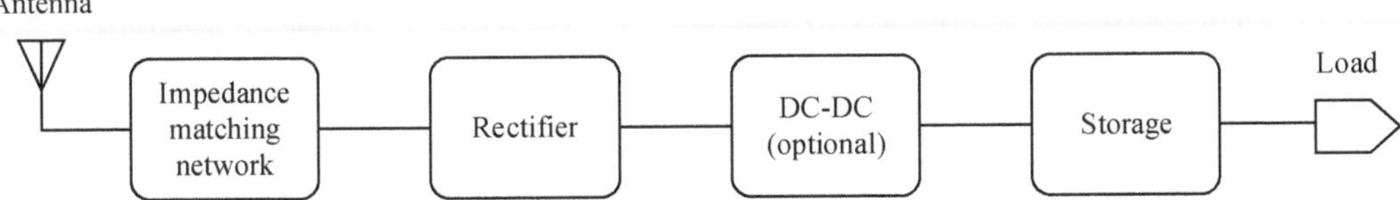

Figure 12. The architecture of the RFEH system.

As the rectifier is a nonlinear circuit, input impedance will vary with the level of $V_{in,rec}$. It is necessary to design the impedance matching network (IMN) to ensure matching between the output impedance of the antenna and the input impedance of the rectifier. In RFEH applications, the IMN has another role, of reducing the reflection from the load to the input; in addition, the IMN sometimes acts as a passive booster to improve the sensitivity of the circuit. The IMN consists of inductors and capacitors in general. Co-design, transformer matching, and reconfigurable matching can also be applied to IMN design [74].

The DC–DC converter can assist in boosting the voltage when the DC voltage output from the rectifier is not enough to power the load. The DC–DC converter also has the role of regulating the load for impedance matching, and is an optional module in RFEH systems. As this DC–DC converter with higher input voltage is not used as the main converter, and the technology is mature, it is not introduced in this paper.

5.1. RF–DC Rectifier

Converters that transform RF signals are called RF–DC, and are measured in terms of power metrics, including sensitivity and PCE. Factors that affect both measures are the input and output signal, the circuit topology, and the device parameters. RF–DC rectifiers are divided into three categories: (1) diode rectification; (2) Dickson method; (3) cross-coupled rectification [74].

Schottky diodes have a low turn-on voltage characteristic [75], and can replace diodes, to improve converter efficiency; however, the manufacturing of Schottky diodes is not compatible with the CMOS process, requires additional steps that can cause increased costs, and cannot be integrated into mainstream CMOS integrated circuits. As CMOS technology grows, diode-connected transistors are gradually replacing Schottky diodes; however, the V_{TH} of diode-connected transistors is still higher than that of Schottky diodes. To address these problems, a series of compensation techniques have been proposed. Static V_{TH} compensation techniques were proposed, to reduce forward losses, and to lower the forward on-resistance to obtain better forward bias; however, a too-low V_{TH} will cause larger reverse leakage current. Active compensation techniques have been introduced, to reduce V_{TH} for forward bias, and to increase V_{TH} for reverse bias, while reducing forward and reverse losses [76–78].

As shown in Figure 13, the cross-coupled differential drive rectifier (CCDR) topology is similar to the cross-coupled design of the CP, except that in the rectifier form there is no clock, but instead a differential-mode signal. The CCDR can reduce the ON resistance in forward bias and the leakage current in reverse bias, and has been widely used because of its low-voltage and auto-switching characteristics: however, large leakage currents are still generated, due to the simultaneous conduction of PMOS and NMOS during the transition phase, which can lead to a narrow-peak PCE, even though the cross-coupled configuration has higher PCE, compared to the Dickson structure.

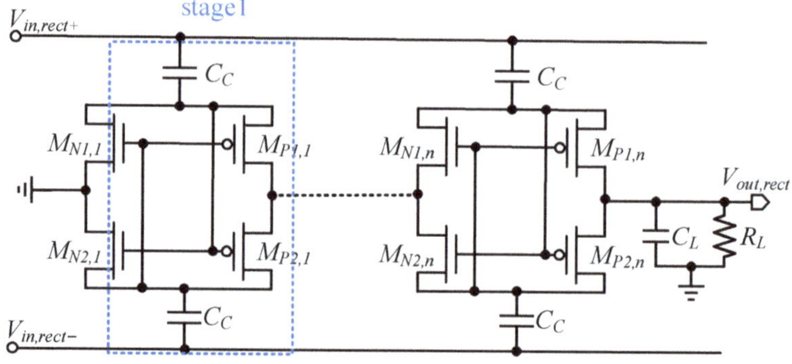

Figure 13. Cross-coupled differential drive rectifier (CCDR) [74].

To widen the operating range of RF–DC converters, reducing leakage current is the key. A self-biasing scheme was proposed in [79], to limit reverse leakage at high-input power levels, which improved the input power range by more than 50%, compared to the conventional cross-coupled rectifier: however, the minimum input RF signal still faced the limiting of the V_{TH}. To compensate the V_{TH}, Gharehbaghi et al. presented a self-calibration technique in UHF rectifiers, while avoiding an efficiency drop due to the wireless power variation [80]: this technique achieved near-constant PCE values under varying conditions, but the maximum PCE was only 34%.

In addition, Khan et al. have combined two rectifiers, and have proposed a reconfigurable power converter using dual path, as shown in Figure 14, switching the series path at low power and the parallel path at high power, to thus maintain high PCE over a wide input power range [81].

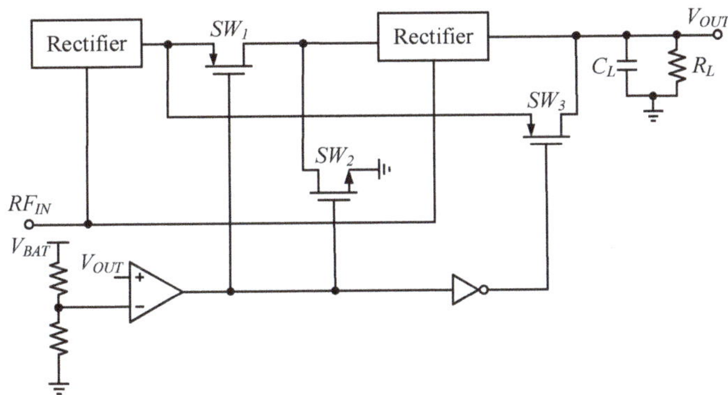

Figure 14. The reconfigurable RF–DC power converter.

5.2. Maximum Power Extraction

Due to the nonlinearity of the rectifier in RF systems, extracting the maximum power from the RF energy can be done using an adaptive impedance matching network method, which can be used to compensate for variations in the input power or load [82], as shown in Figure 15a.

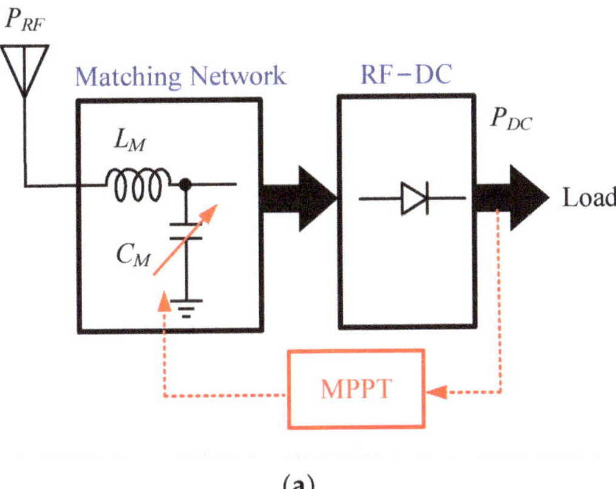

(**a**)

Figure 15. *Cont.*

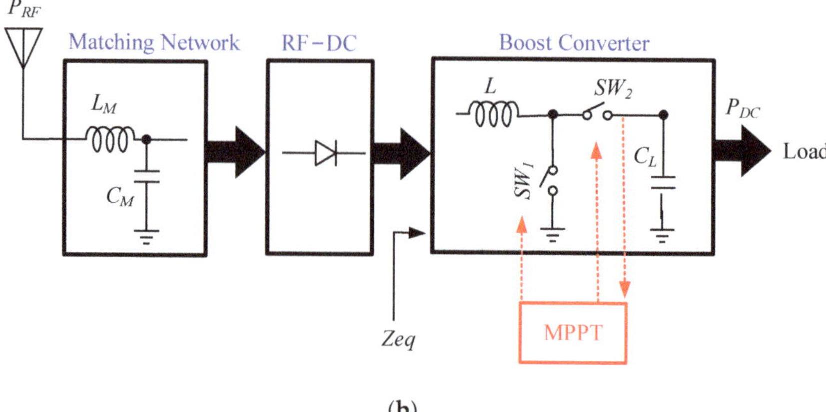

(b)

Figure 15. MPPT for RF–DC: (**a**) with adaptive matching network; (**b**) with cascaded boost converter for adaptive load control [30].

For systems that use the Dickson rectifier, the impedance of the rectifier can be matched to the load by varying the number of the rectifier's stages—namely, the reconfigurable method [83]. If the RF–DC rectifier is followed by a DC–DC converter, the energy transfer can also be maximized by the MPPT method of the DC–DC converter [84], as shown in Figure 15b.

6. Multi-Source Energy Harvesting

To enhance system reliability, energy harvesting systems employing the cooperative collection of multiple energy sources are being increasingly studied.

6.1. DC–AC Hybrid Energy Source Acquisition

The block diagram of a conventional multi-energy harvesting system is shown in Figure 16. All the energy enters the combiner after a single conversion. In addition, the role of an optional DC–DC converter after the combiner is to adjust the output voltage needed by the load.

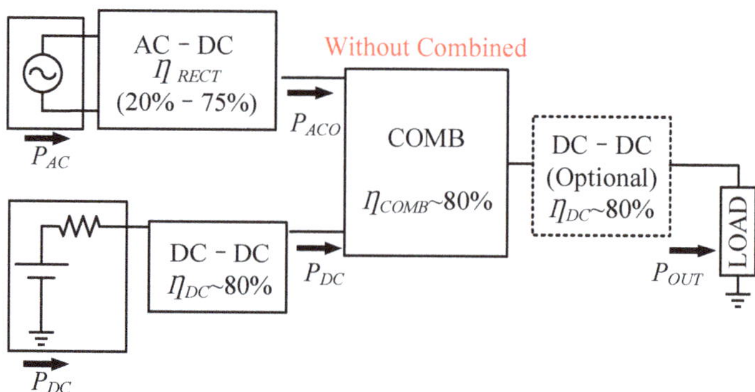

Figure 16. Conventional combiner system structure with mixed AC and DC harvesting.

Combiners and DC–DC converters typically achieve conversion efficiencies higher than 80%, whereas AC–DC converters have efficiencies distributed between 20% and 75%, which are strongly influenced by the input power level and load conditions; the system efficiency is greatly impacted by the rectifier [30].

Liu et al. have proposed a new power combining method, where the AC–DC converter is used as a combiner, and the AC energy is converted only once before it reaches the load [30]. The PCE is improved by combining the energy in the rectification stage, as shown in Figure 17. After combining the functions of the AC–DC converter and the combiner, the system efficiency can reach more than 60%.

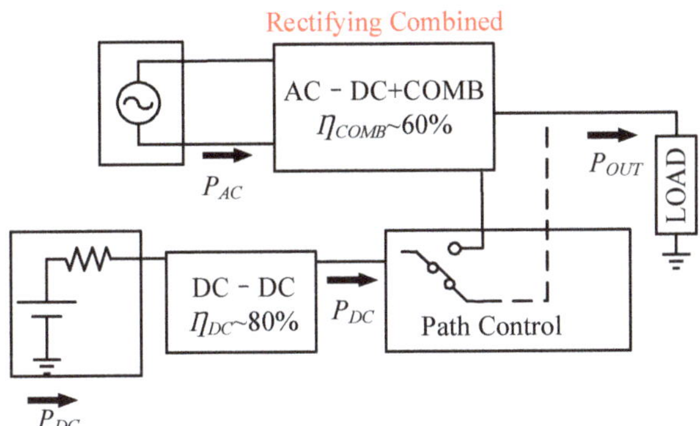

Figure 17. Improved mixed AC and DC harvesting system.

6.2. Architecture of the Combiner

The simplest combined energy structure is shown in Figure 18a, where each energy source is connected to the DC–DC converter through a diode [85,86]: this approach provides a highly modular solution, and allows arbitrary expansion of the input. The drawback is that only the highest energy source can be captured, and the energy from the other ports will be wasted. Another point is that the voltage drop of the diode is also an important factor limiting the efficiency. As in Figure 18b, the use of switches instead of diodes in [35,87] reduces conduction losses, but other issues are not addressed.

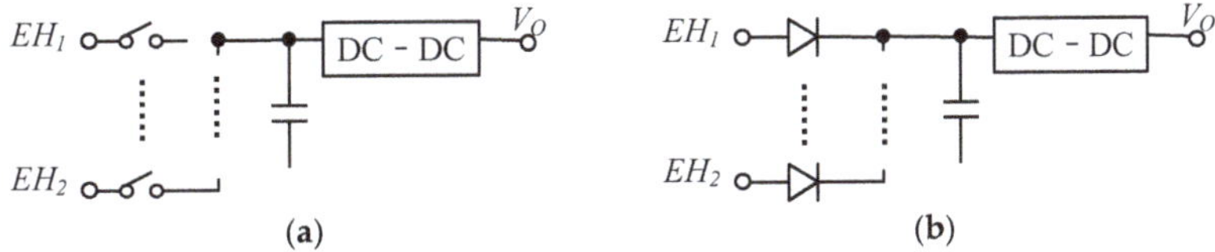

Figure 18. Architecture of energy combining (**a**) diode-connected, (**b**) switch-connected.

The switched-inductor converter structure can be used in combiners, to enable simultaneous multi-source acquisition. A single-inductor multiple-input multiple-output (SIMIMO) converter, with four harvesters, one battery input, and three outputs of 1.8 V, 3.3 V, and one battery, respectively, is proposed in [88] for the MPPT operation, and multiple inputs can be acquired simultaneously. The structure of the inductor-based combiner is shown in Figure 19a.

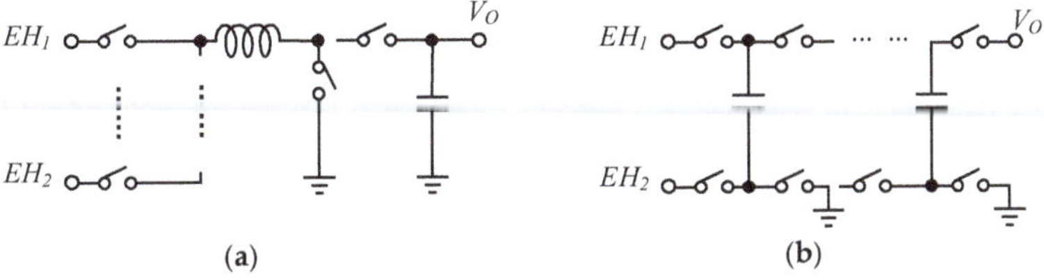

Figure 19. Architecture of energy combining (**a**) inductor-based, (**b**) capacitor-based.

The capacitor-based structure shown in Figure 19b allows for superior integration performance compared to inductor-based. Estrada-López et al. use capacitor-based structure to achieve energy harvesting from a maximum of two input sources with full integration and maximum power output [31]; in addition, the inputs are sorted according to power level, making it possible to automatically select the two inputs with the highest power to be stored.

7. Future Directions and Recommendations

Due to the high energy density and the most mature technology, solar energy is the most widely used. In the design, it is necessary to address the influencing factors that will reduce the system efficiency, such as partial shadows and environmental changes, while finding the balance of system power consumption and MPPT tracking accuracy. Thermal energy harvesting is so low in energy density that the circuit needs to consider self-starting issues and losses in the energy path to be minimized. With respect to DC–DC, the charge pump structure is a promising approach. Based on the charge pump, gate biasing and body biasing focus on lowering the threshold voltage, thus reducing the leakage current, and thereby increasing the efficiency or decreasing the supply voltage, etc. The additional clocking and auxiliary circuitry required to solve these problems increase the overall power consumption of the circuit. The cross-coupled structure can integrate gate bias and body bias techniques to further improve circuit performance, and the advantage of clock boost is the reduction in rise time. Overall, the peak efficiency of the DC–DC converter can reach over 80%. Owing to its high energy density, piezoelectric energy has become the most widely used micro energy, after solar energy, in energy harvesting systems. The efficiency of piezoelectric rectification circuits is extensively measured by the MOPIR; various circuits based on FBR improvements—mainly active rectifier and inductor-based and capacitor-based—have shown in the literature that the MOPIR can reach more than 9. The SSHI is extensively used as a nonlinear method to offset the internal capacitance of the piezoelectric transducer. The advantage of the SSHC is the small area occupied and the possibility of integration. Piezoelectric energy harvesting also requires attention to the effect of different loads on the circuit efficiency. In contrast to other piezoelectric rectification methods, the efficiency of the SECE does not depend on the load. In addition, the design of the piezoelectric energy harvesting circuit also needs to consider the MPPT issue. RF energy input power is weak, and vulnerable to the influence of transmitter distance. The problem of generally narrow peak efficiency also needs to be solved. RF rectification is usually optimized in the CCDR structure, and the peak efficiency can reach more than 75%. Multiple energy co-acquisition is an essential solution for practical application, which needs to be selected with the specific application scenario for suitable micro energy, combined with MPPT technology for energy classification and combination.

Among the MPPT methods applicable to low-power energy harvesting, the P&O method relies on voltage, current sensors, and digital control units. The higher accuracy comes at the cost of higher power consumption. Although FOCV has lower power consumption, it inevitably generates losses during circuit disconnection, and the tracking accuracy is not very high. The recent NN method is only used in high-power systems such as solar energy harvesting, due to slow convergence and high power consumption—however, it has high adaptability to energy sources. To sum up, the more complex but highly accurate MPPT method is suitable for systems with high power levels (mW range), while for systems with low power (μW level), a simple MPPT method with sufficient accuracy may be one of the best implementation options.

WSNs are in sleep mode more than 99% of the time, so their average power consumption is about a few tens of μW, or even less; therefore, when placed in harsh environments, energy management circuits that can efficiently harvest energy from less than 15 μW of ambient energy to power sensors are effective [89]. The miniaturization and off-grid nature of WSN nodes also imposes size requirements on energy management systems. Therefore, the main challenges in the ultra-low-power energy harvesting design scenario include PCE and the circuit size. Reducing the starting voltage is an important means of improving efficiency: the lower the start-up voltage, the more energy can be captured; the higher the efficiency, the more residual energy can be transferred to the load. MPPT is an important way to improve PCE. In the future, it may be possible to allow more complex but effective control methods to be applied to energy harvesting technologies, by optimizing structure and efficiency.

8. Conclusions

This review paper presents a number of schemes for low-power, low-voltage circuits for micro energy harvesting, which indicates that the goal of harvesting ambient energy to supply IoT nodes has been fully developed: however, there are still many challenges to be solved in practical applications. This paper describes the characteristics of the four types of energy, as well as the circuit components, in detail; combined energy, as a method to ensure system reliability, is also presented. Efficiency and circuit size constrain the development of micro energy harvesting: improved efficiency can be achieved by more accurate control circuits and smaller losses; to make the circuit size smaller and easier to integrate on-chip, capacitor-based solutions are more promising than inductors.

Author Contributions: Conceptualization, Q.L. and P.H.; methodology, Q.L. and P.H.; investigation, Q.L. and P.H.; writing—original draft preparation, Q.L. and P.H.; writing—review and editing, N.M.; supervision, N.M.; project administration, N.M. All authors have read and agreed to the published version of the manuscript.

Funding: This research received no external funding.

Data Availability Statement: Not applicable.

Conflicts of Interest: The authors declare no conflict of interest.

References

1. Shabnam, F.; Islam, T.-U.; Saha, S.; Ishraque, H. IoT Based Smart Home Automation and Demand Based Optimum Energy Harvesting and Management Technique. In Proceedings of the 2020 IEEE Region 10 Symposium (TENSYMP), Dhaka, Bangladesh, 5–7 June 2020; pp. 1800–1803.
2. Cabello, D.; Ferro, E.; Pereira-Rial, O.; Martinez-Vazquez, B.; Brea, V.M.; Carrillo, J.M.; Lopez, P. On-Chip Solar Energy Harvester and PMU With Cold Start-Up and Regulated Output Voltage for Biomedical Applications. *IEEE Trans. Circuits Syst. I Regul. Pap.* **2020**, *67*, 1103–1114. [CrossRef]
3. Mayer, P.; Magno, M.; Benini, L. Energy-Positive Activity Recognition—From Kinetic Energy Harvesting to Smart Self-Sustainable Wearable Devices. *IEEE Trans. Biomed. Circuits Syst.* **2021**, *15*, 926–937. [CrossRef] [PubMed]
4. Xiao, W.; Ozog, N.; Dunford, W.G. Topology Study of Photovoltaic Interface for Maximum Power Point Tracking. *IEEE Ind. Electron. Mag.* **2007**, *54*, 1696–1704. [CrossRef]
5. Jeong, J.; Shim, M.; Maeng, J.; Park, I.; Kim, C. An Efficiency-Aware Cooperative Multicharger System for Photovoltaic Energy Harvesting Achieving 14% Efficiency Improvement. *IEEE Trans. Power Electron.* **2020**, *35*, 2253–2256. [CrossRef]
6. Kermadi, M.; Salam, Z.; Ahmed, J.; Berkouk, E.M. An Effective Hybrid Maximum Power Point Tracker of Photovoltaic Arrays for Complex Partial Shading Conditions. *IEEE Trans. Ind. Electron.* **2019**, *66*, 6990–7000. [CrossRef]
7. Jeong, J.; Shim, M.; Maeng, J.; Park, I.; Kim, C. A High-Efficiency Charger with Adaptive Input Ripple MPPT for Low-Power Thermoelectric Energy Harvesting Achieving 21% Efficiency Improvement. *IEEE Trans. Power Electron.* **2020**, *35*, 347–358. [CrossRef]
8. Liu, C.-W.; Lee, H.-H.; Liao, P.-C.; Chen, Y.-L.; Chung, M.-J.; Chen, P.-H. Dual-Source Energy-Harvesting Interface with Cycle-by-Cycle Source Tracking and Adaptive Peak-Inductor-Current Control. *IEEE J. Solid State Circuits* **2018**, *53*, 2741–2750. [CrossRef]
9. Wang, Y.-H.; Huang, Y.-W.; Huang, P.-C.; Chen, H.-J.; Kuo, T.-H. A Single-Inductor Dual-Path Three-Switch Converter with Energy-Recycling Technique for Light Energy Harvesting. *IEEE J. Solid State Circuits* **2016**, *51*, 2716–2728. [CrossRef]
10. Huang, P.-C.; Kuo, T.-H. A Reconfigurable and Extendable Single-Inductor Single-Path Three-Switch Converter for Indoor Photovoltaic Energy Harvesting. *IEEE J. Solid State Circuits* **2020**, *55*, 1998–2008. [CrossRef]
11. Dickson, J. On-chip high-voltage generation in MNOS integrated circuits using an improved voltage multiplier technique. *IEEE J. Solid-state Circuits* **1976**, *11*, 374–378. [CrossRef]
12. Dickson, J. Voltage Multiplier System Using Series Chain of FETs. U.S. Patent US4214174-A, 22 July 1980.
13. Wu, J.-T.; Chang, K.-L. MOS charge pumps for low-voltage operation. *IEEE J. Solid State Circuits* **1998**, *33*, 592–597. [CrossRef]

14. D'Arrigo, S.; Imondi, G.; Santin, G.; Gill, M.; Cleavelin, R.; Spagliccia, S.; Tomassetti, E.; Lin, S.; Nguyen, A.; Shah, P.; et al. A 5 V-only 256 kbit CMOS flash EEPROM. In Proceedings of the IEEE International Solid-State Circuits Conference, 1989 ISSCC. Digest of Technical Papers, New York, NY, USA, 15–17 February 1989; pp. 132–133.

15. Umezawa, A.; Atsumi, S.; Kuriyama, M.; Banba, H.; Imamiya, K.; Naruke, K.; Yamada, S.; Obi, E.; Oshikiri, M.; Suzuki, T.; et al. A 5-V-only operation 0.6- mu m flash EEPROM with row decoder scheme in triple-well structure. *IEEE J. Solid State Circuits* **1992**, *27*, 1540–1546. [CrossRef]

16. Fuketa, H.; O'Uchi, S.-I.; Matsukawa, T. Fully Integrated, 100-mV Minimum Input Voltage Converter with Gate-Boosted Charge Pump Kick-Started by *LC* Oscillator for Energy Harvesting. *IEEE Trans. Circuits Syst. II Express Briefs* **2017**, *64*, 392–396. [CrossRef]

17. Sawada, K.; Sugawara, Y.; Masui, S. An on-chip high-voltage generator circuit for EEPROMs with a power supply voltage below 2 V. In Proceedings of the Digest of Technical Papers., Symposium on VLSI Circuits, Kyoto, Japan, 8–10 June 1995; pp. 75–76.

18. Nakagome, Y.; Kawamoto, Y.; Tanaka, H.; Takeuchi, K.; Kume, E.; Watanabe, Y.; Kaga, T.; Murai, F.; Izawa, R.; Hisamoto, D.; et al. A 1.5 V circuit technology for 64 Mb DRAMs. In Proceedings of the Digest of Technical Papers, 1990 Symposium on VLSI Circuits, Honolulu, HI, USA, 7–9 June 1990; pp. 17–18.

19. Gariboldi, R.; Pulvirenti, F. A monolithic quad line driver for industrial applications. *IEEE J. Solid State Circuits* **1994**, *29*, 957–962. [CrossRef]

20. Luo, Z.; Ker, M.-D.; Cheng, W.-H.; Yen, T.-Y. Regulated Charge Pump with New Clocking Scheme for Smoothing the Charging Current in Low Voltage CMOS Process. *IEEE Trans. Circuits Syst. I Regul. Pap.* **2017**, *64*, 528–536. [CrossRef]

21. Tsuji, Y.; Hirose, T.; Ozaki, T.; Asano, H.; Kuroki, N.; Numa, M. A 0.1–0.6 V input range voltage boost converter with low-leakage driver for low-voltage energy harvesting. In Proceedings of the 2017 24th IEEE International Conference on Electronics, Circuits and Systems (ICECS), Batumi, Georgia, 5–8 December 2017; pp. 502–505.

22. Peng, H.; Tang, N.; Yang, Y.; Heo, D. CMOS Startup Charge Pump with Body Bias and Backward Control for Energy Harvesting Step-Up Converters. *IEEE Trans. Circuits Syst. I Regul. Pap.* **2014**, *61*, 1618–1628. [CrossRef]

23. Chen, P.H.; Ishida, K.; Zhang, X.; Okuma, Y.; Ryu, Y.; Takamiya, M.; Sakurai, T. 0.18-V input charge pump with forward body biasing in startup circuit using 65nm CMOS. In Proceedings of the IEEE Custom Integrated Circuits Conference 2010, San Jose, CA, USA, 19–22 September 2010; pp. 1–4.

24. Kim, J.; Mok, P.K.T.; Kim, C. A 0.15 V Input Energy Harvesting Charge Pump with Dynamic Body Biasing and Adaptive Dead-Time for Efficiency Improvement. *IEEE J. Solid State Circuits* **2015**, *50*, 414–425. [CrossRef]

25. Ballo, A.; Grasso, A.D.; Giustolisi, G.; Palumbo, G. Optimized Charge Pump with Clock Booster for Reduced Rise Time or Silicon Area. *IEEE Trans. Circuits Syst. II Express Briefs* **2019**, *66*, 1977–1981. [CrossRef]

26. Yi, H.; Yin, J.; Mak, P.-I.; Martins, R.P. A 0.032-mm^2 0.15-V Three-Stage Charge-Pump Scheme Using a Differential Bootstrapped Ring-VCO for Energy-Harvesting Applications. *IEEE Trans. Circuits Syst. II Express Briefs* **2018**, *65*, 146–150. [CrossRef]

27. Esram, T.; Chapman, P.L. Comparison of Photovoltaic Array Maximum Power Point Tracking Techniques. *IEEE Trans. Energy Convers.* **2007**, *22*, 439–449. [CrossRef]

28. Han, L. Study of the Micro-Scale Photovoltaic Power System for WSN Nodes. Ph.D. Thesis, Nankai University, Tianjin, China, 2013.

29. Kobayashi, K.; Matsuo, H.; Sekine, Y. A novel optimum operating point tracker of the solar cell power supply system. In Proceedings of the 2004 IEEE 35th Annual Power Electronics Specialists Conference (IEEE Cat. No.04CH37551), Aachen, Germany, 20–25 June 2004; pp. 2147–2151.

30. Liu, Z.; Hsu, Y.-P.; Hella, M.M. A Thermal/RF Hybrid Energy Harvesting System with Rectifying-Combination and Improved Fractional-OCV MPPT Method. *IEEE Trans. Circuits Syst. I Regul. Pap.* **2020**, *67*, 3352–3363. [CrossRef]

31. Estrada-López, J.J.; Abuellil, A.; Reyes, A.C.; Abouzied, M.; Yoon, S.; Sanchez-Sinencio, E. A Fully Integrated Maximum Power Tracking Combiner for Energy Harvesting IoT Applications. *IEEE Trans. Ind. Electron.* **2020**, *67*, 2744–2754. [CrossRef]

32. Mandourarakis, I.; Gogolou, V.; Koutroulis, E.; Siskos, S. Integrated Maximum Power Point Tracking System for Photovoltaic Energy Harvesting Applications. *IEEE Trans. Power Electron.* **2022**, *37*, 9865–9875. [CrossRef]

33. Liu, X.; Sanchez-Sinencio, E. An 86% Efficiency 12 µW Self-Sustaining PV Energy Harvesting System with Hysteresis Regulation and Time-Domain MPPT for IOT Smart Nodes. *IEEE J. Solid State Circuits* **2015**, *50*, 1424–1437. [CrossRef]

34. Jung, W.; Oh, S.; Bang, S.; Lee, Y.; Foo, Z.; Kim, G.; Zhang, Y.; Sylvester, D.; Blaauw, D. An Ultra-Low Power Fully Integrated Energy Harvester Based on Self-Oscillating Switched-Capacitor Voltage Doubler. *IEEE J. Solid State Circuits* **2014**, *49*, 2800–2811. [CrossRef]

35. Rawy, K.; Yoo, T.; Kim, T.T.-H. An 88% Efficiency 0.1–300-µ W Energy Harvesting System With 3-D MPPT Using Switch Width Modulation for IoT Smart Nodes. *IEEE J. Solid State Circuits* **2018**, *53*, 2751–2762. [CrossRef]

36. Newell, D.; Duffy, M. Review of Power Conversion and Energy Management for Low-Power, Low-Voltage Energy Harvesting Powered Wireless Sensors. *IEEE Trans. Power Electron.* **2019**, *34*, 9794–9805. [CrossRef]

37. Lu, C.; Park, S.P.; Raghunathan, V.; Roy, K. Low-Overhead Maximum Power Point Tracking for Micro-Scale Solar Energy Harvesting Systems. In Proceedings of the 2012 25th International Conference on VLSI Design, Hyderabad, India, 7–11 January 2012; pp. 215–220.

38. Wang, Y.; Luo, P.; Zeng, X.; Peng, D.; Li, Z.; Zhang, B. A Neural Network Assistance AMPPT Solar Energy Harvesting System with 89.39% Efficiency and 0.01–0.5% Tracking Errors. *IEEE Trans. Circuits Syst. I Regul. Pap.* **2020**, *67*, 2960–2971. [CrossRef]

39. Bollipo, R.B.; Mikkili, S.; Bonthagorla, P.K. Hybrid, optimization, intelligent and classical PV MPPT techniques: A Review. *CSEE J. Power Energy Syst.* **2021**, *7*, 9–33. [CrossRef]

40. Yap, K.Y.; Sarimuthu, C.R.; Lim, J.M.-Y. Artificial Intelligence Based MPPT Techniques for Solar Power System: A review. *J. Mod. Power Syst. Clean Energy* **2020**, *8*, 1043–1059. [CrossRef]

41. Roy, R.B.; Rokonuzzaman; Amin, N.; Mishu, M.K.; Alahakoon, S.; Rahman, S.; Mithulananthan, N.; Rahman, K.S.; Shakeri, M.; Pasu-puleti, J. A Comparative Performance Analysis of ANN Algorithms for MPPT Energy Harvesting in Solar PV System. *IEEE Access* **2021**, *9*, 102137–102152. [CrossRef]

42. Ahmed, R.; Mohonta, S.C. Comprehensive Analysis of MPPT Techniques using Boost Converter for Solar PV System. In Proceedings of the 2020 2nd International Conference on Sustainable Technologies for Industry 4.0 (STI), Dhaka, Bangladesh, 19–20 December 2020; pp. 1–6.

43. Tabrizi, H.O.; Jayaweera, H.M.P.C.; Muhtaroglu, A. Fully Integrated Autonomous Interface with Maximum Power Point Tracking for Energy Harvesting TEGs With High Power Capacity. *IEEE Trans. Power Electron.* **2020**, *35*, 4905–4914. [CrossRef]

44. Vega, J.; Lezama, J. Design and Implementation of a Thermoelectric Energy Harvester with MPPT Algorithms and Supercapacitor. *IEEE Lat. Am. Trans.* **2021**, *19*, 163–170. [CrossRef]

45. Mateu, L.; Dräger, T.; Mayordomo, I.; Pollak, M. Energy harvesting at the human body. In *Wearable Sensors: Fundamentals, Implementation and Applications*; Academic Press: Amsterdam, The Netherlands, 2014; pp. 235–298.

46. Jhang, J.-J.; Wu, H.-H.; Hsu, T.; Wei, C.-L. Design of a Boost DC–DC Converter With 82-mV Startup Voltage and Fully Built-in Startup Circuits for Harvesting Thermoelectric Energy. *IEEE Solid State Circuits Lett.* **2020**, *3*, 54–57. [CrossRef]

47. Radin, R.L.; Sawan, M.; Galup-Montoro, C.; Schneider, M.C. A 7.5-mV-Input Boost Converter for Thermal Energy Harvesting With 11-mV Self-Startup. *IEEE Trans. Circuits Syst. II Express Briefs* **2020**, *67*, 1379–1383. [CrossRef]

48. Zhang, Z. Research on Distributed Energy Storage Converter Based on Bidirectional Zero Current Switching. Master's Thesis, Beijing Jiaotong University, Beijing, China, 2021.

49. Ghovanloo, M.; Najafi, K. Fully integrated wideband high-current rectifiers for inductively powered devices. *IEEE J. Solid State Circuits* **2004**, *39*, 1976–1984. [CrossRef]

50. Lam, Y.-H.; Ki, W.-H.; Tsui, C.-Y. Integrated Low-Loss CMOS Active Rectifier for Wirelessly Powered Devices. *IEEE Trans. Circuits Syst. II: Analog. Digit. Signal Process.* **2006**, *53*, 1378–1382. [CrossRef]

51. Chang, R.C.-H.; Chen, W.-C.; Liu, L.; Cheng, S.-H. An AC—DC Rectifier with Active and Non-Overlapping Control for Piezoelectric Vibration Energy Harvesting. In Proceedings of the 2020 IEEE International Symposium on Circuits and Systems (ISCAS), 12–14 October 2020.

52. Gasnier, P.; Willemin, J.; Boisseau, S.; Despese, G.; Condemine, C.; Gouvernet, G.; Chaillout, J.-J. An Autonomous Piezoelectric Energy Harvesting IC Based on a Synchronous Multi-Shot Technique. *IEEE J. Solid-State Circuits* **2014**, *49*, 1561–1570. [CrossRef]

53. Kwon, D.; Rincon-Mora, G.A. A single-inductor 0.35 μm CMOS energy-investing piezoelectric harvester. In Proceedings of the 2013 IEEE International Solid-State Circuits Conference Digest of Technical Papers, San Francisco, CA, USA, 17–21 February 2013; pp. 78–79.

54. Du, S.; Jia, Y.; Zhao, C.; Amaratunga, G.A.J.; Seshia, A.A. A Nail-Size Piezoelectric Energy Harvesting System Integrating a MEMS Transducer and a CMOS SSHI Circuit. *IEEE Sensors J.* **2020**, *20*, 277–285. [CrossRef]

55. Du, S.; Seshia, A.A. An Inductorless Bias-Flip Rectifier for Piezoelectric Energy Harvesting. *IEEE J. Solid State Circuits* **2017**, *52*, 2746–2757. [CrossRef]

56. Dell'Anna, F.; Dong, T.; Li, P.; Wen, Y.; Yang, Z.; Casu, M.R.; Azadmehr, M.; Berg, Y. State-of-the-Art Power Management Circuits for Piezoelectric Energy Harvesters. *IEEE Circuits Syst. Mag.* **2018**, *18*, 27–48. [CrossRef]

57. Yue, X.; Du, S. Voltage Flip Efficiency Optimization of SSHC Rectifiers for Piezoelectric Energy Harvesting. In Proceedings of the 2021 IEEE International Symposium on Circuits and Systems (ISCAS), Daegu, Republic of Korea, 22–28 May 2021; pp. 1–5.

58. Wu, W.-L.; Yang, C.-Y.; Wang, D.-A. A Flipping Active-Diode Rectifier for Piezoelectric-Vibration Energy-Harvesting. In Proceedings of the 2020 European Conference on Circuit Theory and Design (ECCTD), Sofia, Bulgaria, 7–10 September 2020; pp. 1–4.

59. Chen, Z.; Law, M.-K.; Mak, P.-I.; Zeng, X.; Martins, R.P. Piezoelectric Energy-Harvesting Interface Using Split-Phase Flipping-Capacitor Rectifier with Capacitor Reuse for Input Power Adaptation. *IEEE J. Solid State Circuits* **2020**, *55*, 2106–2117. [CrossRef]

60. Xu, S.; Ngo, K.D.T.; Nishida, T.; Chung, G.-B.; Sharma, A. Low Frequency Pulsed Resonant Converter for Energy Harvesting. *IEEE Trans. Power Electron.* **2007**, *22*, 63–68. [CrossRef]

61. Li, P.; Wen, Y.; Yin, W.; Wu, H. An Upconversion Management Circuit for Low-Frequency Vibrating Energy Harvesting. *IEEE Trans. Ind. Electron.* **2014**, *61*, 3349–3358. [CrossRef]

62. Ottman, G.K.; Hofmann, H.F.; Lesieutre, G.A. Optimized piezoelectric energy harvesting circuit using step-down converter in discontinuous conduction mode. In Proceedings of the 2002 IEEE 33rd Annual IEEE Power Electronics Specialists Conference. Proceedings (Cat. No.02CH37289), Cairns, QLD, Australia, 23–27 June 2002; pp. 1988–1994.

63. Turner, J.; Ahmed, M.N.; Ha, D.S.; Mattavelli, P. A new approach to the wide bandwidth of piezoelectric transducers for vibration energy harvesting. In Proceedings of the 2012 IEEE 55th International Midwest Symposium on Circuits and Systems (MWSCAS), Boise, ID, USA, 5–8 August 2012; pp. 1064–1067.

64. Chew, Z.J.; Zhu, M. Low power adaptive power management with energy aware interface for wireless sensor nodes powered using piezoelectric energy harvesting. In Proceedings of the 2015 IEEE SENSORS, Busan, Republic of Korea, 1–4 November 2015; pp. 1–4.

65. Shi, G.; Xia, Y.; Xia, H.; Wang, X.; Qian, L.; Chen, Z.; Ye, Y.; Li, Q. An Efficient Power Management Circuit Based on Quasi Maximum Power Point Tracking with Bidirectional Intermittent Adjustment for Vibration Energy Harvesting. *IEEE Trans. Power Electron.* **2019**, *34*, 9671–9685. [CrossRef]

66. Shim, M.; Kim, J.; Jeong, J.; Park, S.; Kim, C. Self-Powered 30 µW to 10 mW Piezoelectric Energy Harvesting System with 9.09 ms/V Maximum Power Point Tracking Time. *IEEE J. Solid State Circuits* **2015**, *50*, 2367–2379. [CrossRef]

67. Kawai, N.; Kushino, Y.; Koizumi, H. MPPT controled piezoelectric energy harvesting circuit using synchronized switch harvesting on inductor. In Proceedings of the IECON 2015—41st Annual Conference of the IEEE Industrial Electronics Society, Yokohama, Japan, 9–12 November 2015; pp. 001121–001126.

68. Fang, S.; Xia, H.; Xia, Y.; Ye, Y.; Shi, G.; Wang, X.; Chen, Z. An Efficient Piezoelectric Energy Harvesting Circuit with Series-SSHI Rectifier and FNOV-MPPT Control Technique. *IEEE Trans. Ind. Electron.* **2021**, *68*, 7146–7155. [CrossRef]

69. Li, S.; Roy, A.; Calhoun, B.H. A Piezoelectric Energy-Harvesting System with Parallel-SSHI Rectifier and Integrated Maximum-Power-Point Tracking. *IEEE Solid State Circuits Lett.* **2019**, *2*, 301–304. [CrossRef]

70. Morel, A.; Quelen, A.; Berlitz, C.A.; Gibus, D.; Gasnier, P.; Badel, A.; Pillonnet, G. 32.2 Self-Tunable Phase-Shifted SECE Piezoelectric Energy-Harvesting IC with a 30 nW MPPT Achieving 446% Energy-Bandwidth Improvement and 94% Efficiency. In Proceedings of the 2020 IEEE International Solid- State Circuits Conference—(ISSCC), San Francisco, CA, USA, 16–20 February 2020; pp. 488–490.

71. Wang, X.; Xia, Y.; Shi, G.; Xia, H.; Chen, M.; Chen, Z.; Ye, Y.; Qian, L. A Novel MPPT Technique Based on the Envelope Extraction Implemented with Passive Components for Piezoelectric Energy Harvesting. *IEEE Trans. Power Electron.* **2021**, *36*, 12685–12693. [CrossRef]

72. Wu, L.; Ha, D.S. An Adjustable-delay SSHC Method for Inductorless Load Regulation for Piezoelectric Energy Harvesting. In Proceedings of the 2021 9th International Symposium on Next Generation Electronics (ISNE), Changsha, China, 9–11 July 2021; pp. 1–4.

73. Costanzo, L.; Schiavo, A.L.; Vitelli, M. Active Interface for Piezoelectric Harvesters Based on Multi-Variable Maximum Power Point Tracking. *IEEE Trans. Circuits Syst. I Regul. Pap.* **2020**, *67*, 2503–2515. [CrossRef]

74. Churchill, K.K.P.; Chong, G.; Ramiah, H.; Ahmad, M.Y.; Rajendran, J. Low-Voltage Capacitive-Based Step-Up DC-DC Converters for RF Energy Harvesting System: A Review. *IEEE Access* **2020**, *8*, 186393–186407. [CrossRef]

75. Karthaus, U.; Fischer, M. Fully integrated passive uhf rfid transponder ic with 16.7-µ minimum rf input power. *IEEE J. Solid State Circuits* **2003**, *38*, 1602–1608. [CrossRef]

76. Kotani, K.; Ito, T. High efficiency CMOS rectifier circuit with self-Vth-cancellation and power regulation functions for UHF RFIDs. In Proceedings of the 2007 IEEE Asian Solid-State Circuits Conference, Jeju, Republic of Korea, 12–14 November 2007; pp. 119–122.

77. Yi, J.; Ki, W.-H.; Tsui, C.-Y. Analysis and Design Strategy of UHF Micro-Power CMOS Rectifiers for Micro-Sensor and RFID Applications. *IEEE Trans. Circuits Syst. I Regul. Pap.* **2007**, *54*, 153–166. [CrossRef]

78. Lim, J.; Cho, H.; Cho, K.; Park, T. High sensitive RF-DC rectifier and ultra low power DC sensing circuit for waking up wireless system. In Proceedings of the 2009 Asia Pacific Microwave Conference, Singapore, 7–10 December 2009; pp. 237–240.

79. Ouda, M.H.; Khalil, W.; Salama, K.N. Self-biased differential rectifier with enhanced dynamic range for wireless powering. *IEEE Trans. Circuits Syst. II Exp. Briefs* **2017**, *64*, 515–519. [CrossRef]

80. Gharehbaghi, K.; Zorlu, O.; Kocer, F.; Kulah, H. Threshold Compensated UHF Rectifier with Local Self-Calibrator. *IEEE Microw. Wirel. Components Lett.* **2017**, *27*, 575–577. [CrossRef]

81. Khan, D.; Oh, S.J.; Shehzad, K.; Basim, M.; Verma, D.; Pu, Y.G.; Lee, M.; Hwang, K.C.; Yang, Y.; Lee, K.-Y. An Efficient Reconfigurable RF-DC Converter with Wide Input Power Range for RF Energy Harvesting. *IEEE Access* **2020**, *8*, 79310–79318. [CrossRef]

82. Liu, Z.; Hsu, Y.-P.; Fahs, B.; Hella, M.M. An RF-DC Converter IC With On-Chip Adaptive Impedance Matching and 307-µW Peak Output Power for Health Monitoring Applications. *IEEE Trans. Very Large Scale Integr. (VLSI) Syst.* **2018**, *26*, 1565–1574. [CrossRef]

83. Abouzied, M.A.; Ravichandran, K.; Sanchez-Sinencio, E. A Fully Integrated Reconfigurable Self-Startup RF Energy-Harvesting System with Storage Capability. *IEEE J. Solid State Circuits* **2017**, *52*, 704–719. [CrossRef]

84. Hsieh, P.-H.; Chou, C.-H.; Chiang, T. An RF Energy Harvester With 44.1% PCE at Input Available Power of -12 dBm. *IEEE Trans. Circuits Syst. I Regul. Pap.* **2015**, *62*, 1528–1537. [CrossRef]

85. Carli, D.; Brunelli, D.; Benini, L.; Ruggeri, M. An effective multi-source energy harvester for low power applications. In Proceedings of the 2011 Design, Automation & Test in Europe, Grenoble, France, 14–18 March 2011; pp. 1–6.

86. Tan, Y.K.; Panda, S.K. Energy Harvesting from Hybrid Indoor Ambient Light and Thermal Energy Sources for Enhanced Performance of Wireless Sensor Nodes. *IEEE Trans. Ind. Electron.* **2011**, *58*, 4424–4435. [CrossRef]

87. Kang, T.; Kim, S.; Hyoung, C.; Kang, S.; Park, K. An Energy Combiner for a Multi-Input Energy-Harvesting System. *IEEE Trans. Circuits Syst. II Express Briefs* **2015**, *62*, 911–915. [CrossRef]

88. Jung, J.-H.; Jung, Y.-H.; Hong, S.-K.; Kwon, O.-K. A High Peak Output Power and High Power Conversion Efficiency SIMIMO Converter Using Optimal on-Time Control and Hybrid Zero Current Switching for Energy Harvesting Systems in IoT Applications. *IEEE Trans. Power Electron.* **2020**, *35*, 8261–8275. [CrossRef]
89. Kawar, S.; Krishnan, S.; Abugharbieh, K. Power Management for Energy Harvesting in IoT—A Brief Review of Requirements and Innovations. In Proceedings of the 2021 IEEE International Midwest Symposium on Circuits and Systems (MWSCAS), Lansing, MI, USA, 9–11 August 2021; pp. 360–364.

MDPI
St. Alban-Anlage 66
4052 Basel
Switzerland
www.mdpi.com

Micromachines Editorial Office
E-mail: micromachines@mdpi.com
www.mdpi.com/journal/micromachines